The Evolution
Explosion

The Evolution Explosion

How Humans Cause Rapid

Evolutionary Change

Stephen R. Palumbi

W. W. Norton & Company • New York London

For information about permission to reproduce selections from this book, write to Permissions,
W. W. Norton & Company, Inc., 500 Fifth Avenue, New York, NY 10110

The text of this book is composed in Sabon with the display set in Centaur
Composition by Tom Ernst
Manufacturing by Haddon Craftsmen Inc.
Book design by Chris Welch
Production manager: Julia Druskin

Library of Congress Cataloging-in-Publication Data

Palumbi, Stephen R.
The evolution explosion : how humans cause rapid evolutionary change / Stephen R. Palumbi.
p. cm.
Includes bibliographical references (p.) and index.
ISBN 0-393-02011-8
1. Evolution (Biology) 2. Nature—Effect of human beings on. 3. Drug resistance in
microorganisms. 4. Pesticide resistance. 5. Breeding. I. Title.

QH371 .P25 2001

576.8—dc21   00-067004

W. W. Norton & Company, Inc., 500 Fifth Avenue, New York, N.Y. 10110
www.wwnorton.com

W. W. Norton & Company Ltd., Castle House, 75/76 Wells Street, London W1T 3QT

1 2 3 4 5 6 7 8 9 0

For my family, my foundation

Contents

Acknowledgments

A writer with his first book is like a beginning harmonica player—dangerously close to exhausting his welcome with family and friends. But I've been extremely fortunate to have a tolerant set of friends and an encouraging family who have always been happy to listen to chapter plans, book titles, jokes, and stories. Mary, Lauren, and Tony have always been supportive as I've filled our home with the mental and physical dross of writing. My parents, brother, and sisters have been avid sounding boards. I also thank the many people who have read through chapter drafts, checking facts and tempering interpretations, including Robert Palumbi, Anne Houde, Barry Sinervo, David Siemens, Dan Griffin, David Haig, Sarah Walters, Mart Gross, Ian Heap, John Doebley, John Losey, John McHugh, Jonathon Losos, Joy Bergelson, Peter Grant, Rosemary Grant, Richard Pollack, Bob Warner, Steve O'Brien, Stuart Levy, Ursula Goodenough, Myra Shulman, David Policansky, Ray Huey, John Endler, Keith Crandall, E. O. Wilson, Jeremy Knowles, and Warren Abrahamson. Bruce Tabashnik was particularly helpful with chapter 6.

Local critics Sarah Cohen, Peter Barschall, Joe Roman, and Krista Ingram were the volunteer fire brigade, helping douse dangerous prose. On the incendiary side was mostly Anthony Palumbi, always pushing for an "interesting" writing style. Eithne O'Brien was a keen first-draft editor and also helped enormously with production. I've learned a great deal about the astonishing world of publishing from Ed Barber at Norton and Ike Williams at Palmer and Dodge. To all of them I am grateful.

The Evolution
Explosion

Chapter One

From the Mountains to the Sea

W̲e waited for the helicopter to find us, listening to it search above for a break in the clouds, thrumming uselessly up there above the fog layer, nearly close enough to touch but not close enough to see. The pilot must have turned and moved away, not knowing he was right on top of us, because we heard the rotors grow quieter as he searched for our small party farther along the misted ridge. Mike Hadfield was on the phone again, a cellular luxury on this Hawaiian mountain peak, calling Honolulu Airport to have the pilot turn back and look again for a place to land.

We'd been snail hunting in the last vestiges of the true Hawaiian cloud forests, searching out the remaining populations of a six-pack of species no one had seen for decades. Unique to the peaks and valleys of the sharp Koolau Range, we hiked along the razor-back ridges, two thousand feet above the warm Pacific Ocean, drenched with mist and blown by winds that hadn't seen land since leaving South America. The snails lived in trees, calmly licking the mildew off leaf undersides and quickly evolving into a riot of different species found nowhere else on Earth. Having finally learned how to raise them in captivity, Mike could save these species from marauding predators like rats and imported cannibal snails, so we had come to find them and escort them from their dangerous ancestral homes. I had scrabbled along with Mike for hours searching for the rare grape-sized globes, and now we were ready for the pickup.

But getting home ourselves might be hard. Mountain slopes rose toward the three-foot-wide ridge like hands pressed together in prayer, fully carpeted by shoulder-high trees dwarfed by the thin soil and whipped by the perpetual wind. Clouds were constantly ripped over the ridgeline by roiling gusts, leaving their cold gift of mist on every surface. The snails loved it, the helicopter didn't.

Twice more the rotors came and went above the clouds, twice more we called Honolulu. We bushwacked to the only flat place we could find, a cropped shoulder of rock on the lee side of the ridge, and when the mists cleared for a moment, the pilot found us. "Impossible to land beneath the ridge," he told Honolulu, and he sat the machine delicately, still flying, straddling the ridge, one hundred sheer feet above us. "Hurry."

We swam up to meet him, breast-stroking through the tree branches like drowning swimmers. Cresting the ridge, we vaulted into the open doors, pushing the delicately wrapped boxes of snails in ahead of us. The helicopter rose a foot, and blew off the mountains toward home.

ANY BIOLOGIST CAN easily find such adventures in Hawaii, where unique species scatter like precious diamonds across the landscape. Thousands of plants and insects, hundreds of ocean species, dozens of birds, and scores of tree snail types—all restricted to the Hawaiian Islands—have fascinated students of the evolutionary process for a century. A riot of evolutionary invention, the species of Hawaii represent some of the most extraordinary examples of how species can change and diversify in the blink of an eye.

But despite this profusion, and our enthusiasm for the native ecosystems of the cloud forests, when Mike and I came down from the mountain and headed home, we passed an even bigger hotbed of evolutionary change, where the rapid evolution of the Hawaiian forests seems glacial in comparison. Nestled on top of a small pass between two old volcanoes—a pass paved by a four-lane highway and choked with commuter traffic—perches a pinkness that represents more evolutionary potential and more evolutionary danger than any seen in the mountains above it. The landmark pink hospital, Tripler Army Medical Center in Honolulu, bustles with modern medicine and coordinated public health, and if you want to find explosive evolution, you should look in a place like this. Here you will find tuberculosis that has resisted treatment, strep throats that have shrugged off erythromycin, and a chilling postoperative infection called methicillin-resistant *Staphylococcus aureus* that can slip into a recovery room and kill after surgery. AIDS patients arrive for checkups, each carrying with them a self-contained arms race between their immune system and HIV, a race that runs on evolutionary fuel, with a grim finish line.

All these powerful diseases are changed by evolution, and all have been reborn recently from the evolutionary cauldron stirred by our technological society. Tripler stands in crowded company— in fact, any hospital in Honolulu or the rest of the country would be as good a bet for finding the products of evolution. No hospital in the world can ignore drug resistance, and drug resistance is evolution come to life.

Farther seaward, toward the glittering Ewa reefs, among the

commercial, agricultural, and military checkerboard called Pearl City, I visited another evolutionary monument, one known throughout the world because something happened there that shook the billion-dollar biotechnology and biopesticide industries. A watercress farmer in Pearl City performed an evolutionary experiment on an introduced pest called the diamondback moth. Thriving in a lush and brilliant green field, pasted between shopping malls and the rising mountain slopes, watercress grows like lightning. But this tasty crop also attracts insect pests. Applying the biological pesticide called Bt toxin year after year to his plants, this farmer accidentally selected a strain of moth never before seen—one more resistant to the deadly Bt toxin than anything else in the world. Almost all other pesticides had already been overcome by resistant insects, but Bt had not. In Pearl City, that record was broken, evolution happened, and the moth won.

SEA LEVEL FINALLY stops my tour, the rhythmic waves bringing foam and sand roiling up Waimanolo Beach to my feet, pulling me gently seaward. Ironwood and coconut shade the sand, but I am not here for long and I want to feel the sun before it slips behind the very mountains I had just descended. This beach, deep with coral sand, consists of irregular shards of calcium carbonate created by tiny coral polyps and anonymous marine algae. The major reef-building coral in Hawaii has been found nowhere else in the world, building up genetic changes over many generations until it was unique, and if we look carefully we can still see this evolutionary clock ticking gently away on the reef.

But we humans have a talent for upping the evolutionary ante and accelerating the evolutionary game, especially among the species that live with us most intimately—our diseases, food, and pests. The same evolutionary process that made this unique coral sand squishing between my toes operates in Pearl City and Tripler and the farms of Kansas. But it is much faster in human hands than on the reef, and for reasons we can see if we look closely, we unleash explosive evolution just by going on with our daily lives, and we don't even know we have lit the fuse.

Chapter Two

Right Before Your Eyes

If I've learned one thing teaching at Harvard, it's that sometimes you have to wear shoes. It doesn't suffice to merely walk the barefoot sands of atolls formed from fragments of unique corals—you have to drag this experience into the lecture hall and light it bright enough to warm the imagination of students whose minds might be trapped in winter ivy. And the best education is the one that bites back, the one that shows with the clarity of glacial ice that the facts and principles of the scientific world are of crucial importance to everyday life.

No one doubts this of chemistry and physics and molecular biology. The average cancer treatment uses all three sciences to discipline the mafia of cells that threatens to kill you through unchecked growth. But my task is to bring home the equally common impact of evolution on daily life—and not through eclectic recourse to scientific theory or historical anecdote. Instead, I need to do it through examples about how evolution in the world around us matters.

The fertile soils of Kansas are part of the everyday life of millions of people—and billions of insects and weeds. And evolution lives among the fields and stalks the checkbooks of struggling farmers—here, like everywhere else, living in the many weed and insect species that have evolved resistance to pesticides. As long ago as 1954, a young Paul Ehrlich, destined to write *The Population Bomb* and scores of other books about the mounting impact of humans on planet Earth, studied the evolution of DDT resistance in flies as a graduate student at the University of Kansas. His work there, and the DDT dustings he and his future wife endured at drive-in movies during Kansas's aborted attempt at mosquito eradication, convinced him of the impact humans were having on the planet.

Local opposition remains strong, however, to teaching the principles of evolution in school. Kansas resident and chemical engineer Joe Smith crystallized this astonishing irony with a few self-assured comments on the 1999 Kansas State Board of Education decision to ignore evolution on tests. "Evolution is just somebody's nice theory," he told *Washington Post* writer Hanna Rosin, "and doesn't impact my life."

Joe is wrong about this. Evolutionary changes that humans did not intend appear so commonly that virtually everyone in the United States and other developed countries has been affected, sometimes critically. The cash-strapped farmer who treats a crop to prevent insect damage must hope that one of the more than 600 species of insects that have evolved resistance does not shrug off his expensive spraying. Simple evolutionary changes like resistance have spawned unassailable insect plagues, ruined crops, and increased farming costs in the United States by more than $1.5 billion a year.

The impact can stalk hospitals too. Any surgical procedure carried out in the past two decades has potentially exposed the patient to bacterial infections insensitive to most common antibiotics. One recent report tallied 30,000 deaths a year in this country due to uncontrollable postoperative bacterial infections—a sign that evolution has moved beyond theory to deadly fact—and recently the Food and Drug Administration has expedited approval of new antibiotics to try to plug this mortal hemorrhage. Viruses are an additional threat. Anyone you know with AIDS harbors a battle within him or her, because in each HIV-infected person, evolution operates so quickly that the virus evolves resistance to new drugs within months.

As we have increased our impact on the planet—by chemical control of pests, using drugs to combat disease, increasing our ability to fashion the physical and biological environment to suit our needs—so too have we become the planet's most potent evolutionary force. Far greater in impact than anything in history, except perhaps the asteroid that wiped out the dinosaurs, we will get the gold medal in the first Olympics to feature planetary upheaval as a sport. We have long recognized the ecological footprint we leave behind on the planet, coached and cajoled by passionate ecologists like Rachel Carson and Paul Erhlich—but much less have we appreciated the depth of the evolutionary footprints that follow.

CHARLES DARWIN FIRST popularized the idea that evolution was powered by natural selection, but in his time selection was not well understood. It seemed a ghost wrapped in thin credibility and was not thought to show itself in real populations or be served at an experimentalist's table. Evolution, after all, generates new species from old, catalogs the passing of the dinosaurs, and scribes the rise from the Cambrian mud of everybody you know. It was supposed to reveal itself, Darwin thought, only by inference and logic. Yet evolution eventually became the focus of the experimenter's art, and once it did, evolutionary change became a commonplace experimental result.

Nowadays, hundreds of experiments confirm the basic evolutionary principles that Darwin painstakingly described, showing why the first major misconception about evolution—that it is just a theory—is wrong. Evolutionary biology has passed onto a new plateau, where experimental rigor and depth of understanding combine to generate rapid insight into the process of evolution. Escorted by new texts and the excitement of insightful experiments, evolutionary biology needs no help from the wishful-thinking side of the scientific menu. Evolution happens, and evolutionary experimenters have learned that they can create a time machine that penetrates the future's gray veil, answering the questions "What will evolve? And why?" without waiting a million years to find out the answer.

These experiments point out the second-most common misconception about evolution: the idea that evolution crawls so slowly that even careful observation will seldom catch it in the act. We now know this crawl can often run, that the evolutionary process can be so rapid that change accumulates while we wait.

The slow evolution that Darwin described could craft changes in species only over the grand sweep of geological time. Darwin espoused slow progress, and was one of its chief practitioners, believing that fossil collections of animals, built up in layered sediments over millions of years, were the only proper setting for observing evolution. Had this been true, then experiments on evolution would be very difficult—as hard as experiments on nuclear fusion—and evolution would remain poorly understood. However, examples of rapid evolution by natural selection abound in our world. Antibiotic resistance, the triumph of HIV over antiviral drugs, size reduction in overexploited fisheries, and resistance of insects to nerve gas pesticides have all happened in the past fifty years, and all have been accelerated by the intensification of ecological change from human activities.

Experimental Evolution

The Other Voyage of the Beagle

Despite thinking that evolution in the natural world was slow, Darwin could see that one particular area of English life in the mid-nineteenth century was actually awash in fast evolutionary change. He found this transformation among the plant and animal breeders of Britain, those organic sculpturers and domestic husbanders who produced many of the farm animals and plants that filled the markets of Europe.

So commonly do animal breeders succeed that we are no longer surprised that domestic turkeys drown in hard rains. It seems natural that they could breed turkeys to taste very good and be extremely stupid. These biodesigners have stocked most of our pantry—oranges without seeds, chickens sporting more breast meat than wings, corn that can't be sprouted—and generated most of our ornamental plants, as well as the pets we don't keep in cages.

They constantly generate new foods too, like the broccolini I had for dinner tonight, a sort of broccoli grown to be tall enough to play basketball. In the sixteenth century, Europeans were inundated with the fruit of the New World, learning to like maize, to plant potatoes, and to trust that tomatoes are not poisonous. Today, our new foods (except for ugly fish mined from the deep sea) tend to be evolved foods, like navel oranges, pink grapefruit, or yellow peppers. These alterations, as much a part of industrial food development as new ways to store Twinkies, spring from selective breeding—artificial selection—of agricultural species.

The triumph of the domesticator's art, even before navel oranges, derives from careful artificial selection leading to fundamental changes in the appearance of the species in question. In fact, domesticated animals and plants were the first sophisticated tools the human race produced, and they represent the only prehistoric technology still widely used today. These selective endeavors represent uncontrolled experiments in evolution—the passionate canvases of a biological artist—and they were the first breathless hint that the basic logic of the evolutionary process was sound. Dog breeds prove easy examples of the evolutionary process, crystallized in huskies, retrievers, and wiener dogs that go down holes after prey. And so, a rheumy-eyed beagle hound represents the purebred climax of an evolutionary voyage predating Darwin's own evolution on a ship of the same name.

Flies in the Face of Fact

> We dedicate this bit of real earth, its sprouting plants and its breeding animals, here and now to the study of the laws of the evolution of organic beings.

So announced Charles Davenport on June 11, 1904, culminating the dedication of the Station for Experimental Evolution, in the former whaling town of Cold Spring Harbor, New York. Gregor Mendel's laws of genetics had been rediscovered but a few years before, and Davenport was enthusiastically convinced that the study of heredity and breeding would crystallize evolutionary thinking and revolutionize biology. The laboratory he started, which eventually grew into a biological powerhouse, began auspiciously with a firm focus on Darwinism.

On the staff of the new station was Edwin Carleton Mac-Dowell, a careful investigator of evolution in the physical features of animal populations and the power of selection to drive it. For a subject, he pragmatically chose a tiny fruit fly that had become popular with the proponents of the new scientific discipline of genetics. The fly was complicated enough to serve as a good model for how evolution could work, was easily manipulated, and could be grown in small bottles. At the fast pace of fly life, fifteen or so generations might be crammed into a single year, enough for a longer family tree to grow during a college semester than the Pilgrims produced between 1620 and 1776.

MacDowell conducted a series of simple but powerful experiments in which tiny features of the flies' bodies, such as the small bristles on the thorax, were used as the palette of experimental evolution. The fly thorax, which emerges under a microscope as the middle section between the tiny head and the grublike abdomen, is studded with a small crop of short cuticular hairs. From an initial population of wild flies, MacDowell grew a generation in the lab and painstakingly counted the bristles on each one's thorax. In one line, he chose only the flies with the largest number of bristles to start the next generation. The rest were discarded. To form another family line, MacDowell chose only the flies with the smallest number of bristles. Generation after generation, MacDowell grew the flies to adulthood, and chose which would survive and breed, keeping track of the number of bristles in all flies in the bottles, discarding losers and lucky breeders alike, and comparing the averages over time.

The result of this experiment, published in 1915 in *The Journal of Experimental Zoology*, was probably not a surprise to chicken farmers, who had known for years about the power of artificial selection. The experiment showed a steady and significant increase over eight generations in bristle number in the hairy fly line and a stunning decline in the naked fly line. By choosing which flies would breed, MacDowell had generated what Darwin called artificial selection. By continuing the experiment over many generations, MacDowell showed that evolution could happen in the fly populations, changing their features and those of their descendants.

So what? This question reverberates frequently in science— mostly asked by other scientists, or the mothers of graduate students. So what did we learn from an experiment on fly bristles? Do these bristles matter, even to the fly? Probably not, for they play no critical role in a fly's life. Wasn't this what animal breeders had known all along? Yes. But at another level, the results are not just interesting but revolutionary, not just important, but pivotal. The experiment helped open evolution to critical, objective testing and provided the experimental platform that supports evolutionary experiments to this day.

ANYONE CAN COAX a simple sound out of a piece of wood, a sharp cellulose bark or a deep tribal thrumming. It's simple to pluck a trembling string and get a long, Zen-like tone, a meditative frequency filling the air. But a violin raises these simple acts to the level of near magic.

The violin maker, while craftsman and artist, is not the musician. The handcrafted beauty the maker produced lives not only in the lines and the lacquer and the ebony pins. The curves and the luster may be a lyric in themselves but they are not the final chorus. The beauty of the violin lies tightly wound in its potential to produce music, and the promise of tomorrow's song. Like a prima donna's parents, the violin nurtures the sound; it lets music fly only when fully formed and beautiful. The violin maker's art triumphs when it creates this machine of creation, producing a crucible of imaginative energy that can husband sounds the maker never experienced, that the player only dreamed. A new musical instrument is a promise with all the potential of a nestling canary, one with a future as unknown as the dreams of a sleeping infant.

In science, a single experiment is a fiddle tune. It can be short and lively, or long and serious. It can amuse temporarily, or it can take over your brain, spending days rolling around inside your head. But if an experiment is a song, then the way we conduct the experiment is the violin itself: the method used to do the research—the way we produce the result—is the instrument on which the experiment plays. Sometimes the method is a technology, a tool belt of sunlight and prisms. Sometimes it is just a way of asking a question, like Mendel's meditative experiments in the garden of peas. And the first answers squeak like the first notes from a new violin—amazing in themselves, but even more amazingly pregnant with the promise of future melodies.

The first experiments on evolution were squeaky notes of a future symphony. They proved the general features of evolution as an experimental science—that evolution experiments could work over a time frame that people could observe, and that evolution happened under conditions that could be experimentally altered and therefore understood in minute detail. Biology was letting us pick up evolution's violin, and we were well on our way to tuning the strings.

An Order of Curly Flies

The first rule of a stage magician is never to repeat a trick, never to let the audience have another chance to see the sleight of hand or where the rabbit really came from. This gives the magician an edge over the audience, a power to control the atmosphere of disbelief and to create the different reality that a magic trick requires.

The distinct power of science lies in the opposite philosophy: you can battle skepticism through judicious repetition and the lesson of honest openness. Take the audience home and show them over and over the way the experiment works, describing how the thing was done in such detail that an army of imitators can strive for the same answer. Repetition is the hallmark of a good experiment, the scientific Good Housekeeping seal of approval, and builds confidence in a skeptical audience.

Laboratory settings lend themselves to repetitive work because conditions can be controlled and details can be replicated precisely. Over and over the experiments can be run, over and over the results can be tallied, until consistent answers emerge. Like most crucial experiments, fly evolution studies have been done this way: choose hairy flies and hairy flies become more common. Because of repetition, we do not interpret the bristle experiment as a fluke.

Another kind of repetition makes scientific sense too. Sometimes an experiment should be altered slightly as it's repeated. Can other physical features besides hairiness evolve? One fly experiment selected for curly wings by laying the flies flat on a tabletop and lowering a sticky board over them. The curliest wings stuck to the board first. When the board was lifted up, it brought away these curliest winged flies, little fly legs kicking in the chill lab air. The irritated flies were unstuck and were then released into a fly nirvana of goopy food, where they quickly forgot their pique and bred the next generation. Selection soon produced a line curlier than a L'il Orphan Annie hairdo, with wings that were useless for flying.

More basic traits—those more fundamental to the organism's survival—have also been shown to evolve. Body size at birth has been artificially selected in mice, and it can be increased or decreased over generations of arduous biological bookkeeping. Initial results were fast: over the first few generations, body size at birth dropped in the lines selected for small size, and increased when selected upward. But these changes quickly leveled out, and despite continued selection, body size did not drop below a particular floor or rise above a particular ceiling. The reason was soon clear: the artificial selection imposed by researchers was in turn opposed by natural selection.

Unlike fly bristles, body size at birth matters a great deal—too small a baby mouse will not survive, even in the artificially benign environment of the lab, so artificial selection can not drop mouse size below that point. On the other extreme, giving birth to too large a baby mouse becomes a threat to Mom either because of birth complications or punishing food demands; as a result, natural selection discourages large size at birth too.

This mouse experiment showed that evolution could happen in a mammal (closer to home than flies) and that something critically important like birth weight could be altered by artificial selection. Furthermore, the experiment demonstrated that the power of artificial selection was limited by thresholds built into the basic body plan of a mouse. Too small or too large: both strategies were doomed to produce mice unfit for even the marshmallow life of a lab animal.

Just an Act

Aggressive and solitary, foxes are ranched in Siberia for their fur production. Throughout the 1970s and 1980s, in an unprecedented experiment in social evolution, which had unforeseen consequences, foxes at one ranch were carefully bred for being tame.

Unlike Antoine de Saint-Exupéry's Little Prince, the investigators did not want to tame an individual fox, they wanted to breed tame foxes without actually taming a single suspicious canine. Over fifteen generations, only those foxes most tolerant of humans were used in the breeding program. At first, these foxes interacted poorly with humans. On a 5-point "tameness" scale, they were only 1s.

Within a few generations, however, 2s and 3s were common. By the tenth generation, the animals seemed to enjoy human contact and were social to an extent never seen in offspring of wild animals. By the last generation, some other traits began to appear—in particular, barking (fox puppies bark, but adults do not). So it seems that by selecting for tameness, the Siberian experimentalists were selecting for a whole wardrobe of characteristics linked with tameness and associated with juvenile behavior patterns. Tails held high instead of low, ears perked forward, playfulness, and barking—all seemed to be signs of foxes that behaviorally were hanging on to their youth far longer than usual. Behavioral selection had led this population to evolve, in ways that showed how the word "tame" was inadequate and perhaps misleading. Evolution had happened in response to selection for tameness, but the response was deeper than expected because it is impossible to select for only a single trait when whole organisms are chosen to be parents of the next generation.

IN ORDER FOR experimental evolution to work, experimenters must be able to select the desired traits by picking and choosing parents for the next generation. In general, this approach has been very successful and has been used in flies to increase hairs, life span, body length, wing venation, and sexual apathy. It has been used to alter the physiology, morphology, or behavior of insects, birds, fish, mammals, worms, bacteria, and a few viruses.

So commonly does laboratory-applied selection change a population that failures are fascinating; these exceptions generally reveal a basic but unforeseen property of organismic plumbing—or an investigator's goof.

One fly experiment, which tried to select for increased ability to tolerate temperature extremes, consistently failed. The selection device—a long, vertical, heated tube that increased its temperature gradually—was used to warm the flies in order to test their ability to withstand temperature stress. Flies were released onto the tube, and milled around waiting for something to happen. As the tube got hotter and hotter, the flies started falling off—physiologists call this heat-induced torpor, but swooning could describe it too. Wimp flies fell first and the really tough, walk-on-coals flies made it further into the experiment before their plunge.

The device worked quite well, and the investigators collected the flies that fell first to make a line of flies intolerant of high temperatures. Other flies, having fallen once the tube heated to a particular point, were collected later. Still other flies never fell, and were collected as they clung to the tube after the experiment ended. Later generations were bred from this clinging group, and offspring were tested in tiny fly saunas. There they proved to be more active at high temperatures than their ancestors. However, their tolerance to heat, measured by the temperature at which the flies passed out, did not increase over the generations.

What was wrong? Artificial selection drives evolution in features other than those that biologists expect. The Siberian foxes selected for tameness begin to bark, not because the investigators chose this trait but because this trait emerged accidentally when an overall tameness index was invented. Similarly, our problem flies were *not* being selected for their ability to remain active at high temperatures. Instead, the tube device had had an unexpected effect: it selected for flies that would still hang on to the tube after passing out, using tiny foot claws that dug in and suspended them above the pit of fly failure. The swooned but clinging flies were collected by the experimenters and used to breed future generations of supposedly heat-tolerant flies, but what really occurred was artificial selection for the ability to hang on while incapacitated. True, the experiment taught an evolutionary lesson, but the nature of the selection differed from the one originally sought. The selection process, blind to the goals of the experiment, produced a different evolutionary result.

Why Opposites Don't Attract

Late in 1995, a controversial study was published by the Royal Society of London. It covered a subject that would have repulsed the society members of Darwin's day—whether female college students preferred the smell of particular male students based on their genetics. Stored in Tupperware containers, T-shirts worn by unwashed males were sniffed by brave female testers and graded on a preference scale, after which all the participants were scored for the genes they possessed at major immunological loci. The women in the study preferred some shirts over others, often those worn by males who differed from them most strongly at the genes assayed.

The study was driven by a widespread biological observation: strong mate choice is a pervasive feature of life, typically keeping some individuals from mating even though they could produce perfectly healthy offspring. For simple marine creatures without brains as well as for mammals with complex pre-mating rituals involving expensive designer clothes, mating clues help determine mate choice within species. From sea urchins in a tide pool to com muters in a car pool, individuals choose among potential mates, generally favoring only a few of the possible candidates.

Because mate choices vary between individuals—like in the T-shirt example—could artificial selection for mate-choice patterns lead to mate-choice evolution? And could mate choice become so different that reproductively segregated populations were the result? Such questions address whether evolution by natural selection can lead to the generation of new species, the main tenet of Darwin's *Origin of Species*.

Over and over, experiments on fruit flies have shown that divergent selection can lead to two fly lines that prefer to mate within their own line when brought back together. In the 1950s and 1960s, a battery of experiments subjected flies to divergent conditions: more bristles versus fewer; warm, moist versus cool, dry conditions; a tendency to walk up a vertical plane versus a tendency to walk down. In most cases, the artificial selection succeeded in changing the way the flies looked or behaved, and also had the secondary effect of producing strong mate choice among flies that were similar. One study selected on differences in activity level, choosing flies that were more active in one line and picking couch potato flies in the other. After forty-five generations of selection, the lines were brought back together, and the investigators found the highly active flies tended to mate together, as did the slower flies. By contrast, matings between activity levels were rare.

Bill Rice at the University of California at Santa Barbara must have been trapped in a fun house while young, because he designed the most elaborate of all selection devices for separating flies; it involved a labyrinth of tunnels and ramps and heat sources and vapor trails, all meant to divide a confused population of flies into two groups that differed in their habitat choice as strongly as possible. Because of the long series of obstacles that had to be overcome by the flies if they were to be successful, Rice's experiments generated two groups of very different individuals, each of which consistently made different habitat choices. And the flies made different mate choices too, tending to mate with other flies that had the same habitat preferences. Not only did these experiments show that evolution could act on multiple behaviors at once, but they also showed that such divergent habitats could generate populations of flies that chose not to mate with one another.

Selection in the Wild World

A lab concentrates wonder like a biography concentrates a lifetime, with all the best mysteries hung in chapters for study and storage. By design, the unusual becomes commonplace, and events unfold like high school love notes, quickly, furtively, and maybe only once. What's normal inside—superconductivity, animal cloning, and graphs of hairy flies—just seems a peculiar story outside. By and large, labs are not a normal place to work.

Therefore, evolution in the lab represents a great scientific tool, an instrument of discovery that can write symphonies. But does the music of evolution play in the natural world? Do populations not bottled or caged or tracked by a microscope evolve under selection, artificial or natural? Yes, they do.

"JESSE JAMES" WAS how I thought I heard the man introduce himself, sitting astride his compact horse, looking at me with a predatory grin. But I blinked and looked again, and the Western desperado image dissolved—Jesse Jones sat simply and comfortably dressed in L. L. Bean flannel, sporting a riding helmet and UV-coated sunglasses. True, he commanded a brown and spirited mustang, and occupied his handmade saddle with the intense pride of craftsmanship, but he was a modern rider not a bank-robbing legend. Amid the rolling, grassy hills of a Maryland piedmont farm, I admired Jesse's horse and waited for my daughter and her cousin to emerge from the small, muddy stable with horses of their own. Because everyone was so casual on this early spring afternoon, when idle chat easily laid claim to the day, it was some surprise to be treated to an impromptu lesson in horse evolution.

"This horse has no shoes—you don't shoe a mustang except for special times," Jesse explained.

He leaned over the neck of his horse, "Spanish bloodlines in her," he confided, as if he was keeping this a secret from the horse, "and the best you can find."

"No mustangs need shoes?" I asked, thinking maybe this was a case of artificial breeding. "The Spanish blood?"

"No." He denied the Spanish connection. "These lines were released out west and ran free there, but in west Texas, they had to fight off the mountain lions and what all. West Texas stock evolved to run fast and far, and have hooves that are hard. That's why they don't need shoes—because of being run after, across the plains."

"Must have been a fast change," I said.

"Yep," Jesse agreed. "Only way to stay alive in west Texas."

I KNOW OF no one who could have measured the evolution of the mustangs during their time in west Texas. There are no data on shifts in hoof thickness or hardness or growth rate, no observation of predation and selection in a feral population of these muscular and magnificent horses. Such data might have gone down in scientific history as the first observation of the evolutionary process in a natural setting, the direct proof that Darwin was right about the impact of selection and variation on the way species looked and behaved and lived their lives. In fact, demonstrations of evolution in wild populations were once as uncommon as boots at the beach. However, once people started looking, evolution in the wild—like evolution in laboratory settings—was very easy to see with their own eyes.

Gallons of Guppies

Henry Huggins was the hero, as I recall, of a book I borrowed from the library when I was eight or so. In a chapter called "Gallons of Guppies," two scenes stick firmly in my mind. In the first, Henry visits a fish store and wonders how long fish can live in little plastic bags hung on the wall. Fish don't live in bags, it turns out (an early lesson in aquaculture), and Henry needs a fishbowl too. In the second scene, as Henry was admiring his new pet guppies swimming contentedly in their fragile global home, he had male guppies, female ones, and . . . specks. The specks were a surprise, but Henry determined upon close inspection that they were tiny guppy replicas—thin little baby fish that are born, alive and already hungry, from a mother guppy pregnant nearly 100 percent of the time.

Henry, a good guppy grower, soon had many globes, bottles, and jars—and eventually everything he could find—filled with guppies. All the guppies were swimming and birthing and growing and breeding, all taking up his time and needing more attention than Henry could give.

It was a nice story—and pretty plausible, given the way guppies pump out offspring every month or so, and then become grand-guppies a month later when their first batch of offspring begins life's great multiplication. But of course, because it was written for kids, the story left out a critical and fascinating part. Specks have mothers, but they have fathers too, and how mothers and fathers get together certainly wasn't part of the juvenile literature of the 1960s—when even Ozzie and Harriet had single beds. So Henry ignored the males pretty much (beyond mentioning how colorful they were) and concentrated on the much safer miracle of birth, especially how frequently miracles could be had when guppies were around.

Henry never stopped to wonder why only the males are pretty. The females are dull green-brown, a comfortably fishy color that resembles a paint mixer's gloomy day. The males though have a rainbow flash of color splashes, a panache of blue, red, orange, and green that cloaks their singles'-bar manners and writes their love sonnets in scales of marvelous hue. The females prefer to mate with brightly colored males, so male color is a device for guppy love, and Henry missed it all.

The female's dull color actually has a lot of practical value. Guppies, usually gentle creatures, do not win arguments with large predatory fish; they rely on stealth to survive. They live in small streams and learn to avoid the neighbors that would eat them. Predatory fish hunt by sight, attracted by the flash of color. The courting costumes of males, then, can help them find mates but it exacts an awful price: attracting predators instead of a potential mate.

SELECTION ACTS ON guppy coloration in two different ways. Predatory fish find and eat the most colorful males, and so the dullest ones survive the longest. However, because females prefer to mate with the brightest males (we are talking about color here, not Einstein guppies), in the absence of predators, colorful males have the most offspring. Darwin called this special type of selection *sexual selection* and hypothesized that such systems of mate choice explain many of the elaborately different features of males and females of the same species.

Experiments in artificial ponds and streams reveal how color evolves in guppy populations. Without predators around, females choose the most colorful males. After a few generations of selection, the average male displays a veritable rainbow compared with his forebears; he has become a harlequin gigolo with a Casanova attitude and the power to beguile.

But place a few predators in the tank, and things change rapidly. All of a sudden the dull-colored, less aggressive males quickly come to dominate. Females still mate preferentially with the few brightly colored males, just before lunch, but these males do not last as long, and the lifetime reproductive success of dull males outscores the bright ones.

The tug of different kinds of selection also determines guppy fashion in real streams. From the highlands of Trinidad flows a handful of meandering streams, splashing through sheltered forest galleries to fall into downslope pools and busy rivulets. The University of California's John Endler, who has an eye for a good guppy, has splashed these streams to understand evolution and has recorded guppy evolution in action. The lower streams are tough neighborhoods with big-fish, little-fish rules. Predators stalk there with silent fins and lawyer's teeth, and most males are dull—or young.

Upstream, the hissing waterfalls paint a police line that big fish do not cross, and the headwaters remain relatively free of fish predators. As predicted, and as shown by Endler's careful experimental studies, male colors are brighter upstream than downstream. What's more, when Endler introduced predatory fish upstream, he saw guppy male coloration there quickly evolve to be much duller. These experiments, inside the lab and out, show that fish coloration changes like human fashions, responding to the balance of attraction and danger, and evolving quickly in each stream to reflect the local mix of females and predators.

Thanks for the Sparrows

On an inauspicious day in 1851, in Brooklyn, a bird lover entered the history books by releasing a population of sparrows brought from England expressly to populate North America with their chipper bright eyes. I wonder what the native birds—the orioles preening their tiger splash of orange and black feathers or the buntings in their battle for the most brilliant blue—thought when such nondescript browns and blacks were so purposefully added to the woods? Whatever their attitude, English sparrows found North America to be a good neighborhood, and took up residence across the continent. From New York to the West Coast, from Alberta to Houston, sparrows multiplied and spread, set up housekeeping, bred new broods, and moved on. Sparrows have settled into most of North America as if they belonged here.

Therefore, it comes as a surprise, given the recent history of sparrows and their origination from so few ancestors, that a good birder can tell where a sparrow comes from by just looking at it. Sparrows from the frozen north weigh in a bit bigger; this is presumably an evolutionary shift favoring chunky birds that can stay warmer through the winter. Sparrows from the south have a yellowish tinge, perhaps allowing better camouflage in the vegetation they call home. These differences should not be overstated—nobody thinks these different birds are different species, and some of these differences may play no role in survival at all—but where there was once only a single flavor of this sparrow, now several take to the skies.

Such differences in sparrows from place to place have one common cause—evolution. As they moved into new areas—leaving the hot and humid summers of New York for the icy winters of Alberta—natural selection began to change a few of the features that help determine which birds survive best. Part of that process was caught and recorded by Hermon Bumpus on February 1, 1898. Bumpus was a new faculty member of Brown University in Providence, Rhode Island, on his way to eventually being the director of the American Museum of Natural History in New York and then the fifth president of Tufts University in Boston. It had been a harsh winter in the Northeast and Bumpus found a small sparrow flock that had been knocked to its feet by a passing storm of "snow, rain and sleet." Death claimed about half the birds. Bumpus measured both the survivors and the deceased.

Death from the storm did not strike randomly, Bumpus concluded, in a now classic example of natural selection in action. Instead, the slightly shorter and lighter birds had a somewhat better chance of survival. Bumpus described slight differences between survivors and the deceased in most of the bird bits he measured: legs and wings, brains and beaks, height and mass. Further inspection of the data revealed an even stranger pattern: shorter birds tended to survive better, but the shortest bird died. In fact, birds that were extreme in some measurement—the largest or smallest, the shortest or tallest—tended to be in the group that perished. Bumpus lamented over one deceased bird that had some albino feathers, "already cursed by six [other] abnormalities, the most miserable individual in the entire collection." In stark contrast, the surviving birds tended to be those sporting few extreme features. Following this evidence, Bumpus concluded that there must be an ideal sparrow type for the local climate, and that birds deviating most from the ideal had the least chance of survival. Natural selection stalked these birds, and evolution happened.

Darwin's Mistake

In his early travels and famous observations, Darwin was strongly motivated by geological theories that were changing nineteenth-century science. Promoted chiefly by geology pioneer Charles Lyell, who cataloged the geological formations of England in mid-century, and echoed by many others, these theories explained the many hued layers of rock that make up the sides of steep gorges and chalky cliffs scattered across the English countryside. Lyell thought these layers had developed over slow, stone years through the gradual depositing of different silts and sands and the layered soils of failed continents. As current science had it, the Earth was not old enough for this centuries-long epic poem of geological growth. If Lyell was right, the Earth had to be hundreds of millions of years older than anyone had thought.

Having read Lyell's persuasive *Principles of Geology* while aboard the *Beagle*, Darwin grabbed this extra time and had it do double duty by allowing natural selection to slowly alter the features of a species, and to slowly build up the differences needed to call an altered population a new species. It didn't happen very quickly, Darwin thought; evolution by natural selection had had a long, long time in which to operate.

Now, when I was a graduate student in Seattle, a professor there had a rubber stamp that came out late at night when even the most dedicated dissertation writers had gone home—or maybe when everybody else was at the pub arguing over beer and shelled peanuts about ecology and hypothesis testing. When the building was quiet, the stamp came out, striking repeatedly all over everything in red ink. Manuscript pages, posters, new graphs of old results, copies of seminar papers, grant proposals, and nearly every written product of nearly every student and faculty member was smitten by the secret stamp. We all returned to read on our papers and proposals that "Darwin already thought of it."

So sad, because this was so often true. Darwin sat for decades pondering the clues of ecology and evolution, and buried in the labyrinthine prose of his thousands of pages were a university of ideas that would later be discovered again. Darwin named the three kinds of selection: natural, artificial, and sexual. Nobody in the 150 years since has discovered a fourth. Darwin discovered the principles of ecological interaction and fathomed the impact of complicated mating behaviors on the ornamentation of male birds and antlered mammals. He knew about the importance of earthworms to soil fertility, and that barnacles had legs. He had seen the truth about cave animals—that they were most usually related to the creatures living just outside the mouths of their own caves, not to other cave creatures around the world—and he knew where the largest mountains of the world had disappeared to.

But he made a serious error about the speed with which evolution can happen. The insulated sparrows of Alberta or the guppy in his coat of many colors testify to the speed of evolution in the real world. And then we have the hairy flies, with their curly wings, and all the other sports of the experimentalist's craft. They show that evolution happens, and they show that it happens very quickly.

A Bird Named Darwin

Darwin's ideas were slow but his finches, which are among the best studied examples of evolution in the wild, are fast. Small brownish birds, Darwin's finches are nondescript, like the conservative side of a professor's wardrobe. They get their name from the studies Darwin did of the variation between species. In that way, they constitute a monument to the importance of Darwin's experiences in the Galápagos Islands during his voyage aboard the *Beagle*. As they did when Darwin saw them, the birds hop and helicopter through the low vegetation of the dry Galápagos, searching for seeds in dry weather, searching for mates when the rains come and the food bounty increases.

Seeds can be difficult food. Imagine a sit-down dinner of Brazil nuts and macadamias. There's the knife and fork, but where's the hammer? A vise grips would be placed next to the bread plate in a really good restaurant, but in the Galápagos, the birds must rely on their natural toolbox of claw and beak. A variety of different tools are needed to handle the different-sized seeds that Darwin's finches encounter. Large beaks can break into tough large seeds, but they are clumsy tools for the delicate handling of smaller softer items. Small herbs produce seeds so small that a thousand would weigh only an ounce. Other, tough plants produce dried horrors an inch wide and so difficult to eat that only starving birds will try.

Although such environmental variation could be documented in countless habitats for countless species, these particular seeds and finches have captured the attention of a band of biologists that have pitched their careers like field tents on the Galápagos peaks.

Led by Princeton scientists Rosemary and Peter Grant, they track the environmental changes of these birds as avidly as preelection pollsters and have discovered through their careful intuition that evolution can dance a quick tune in the Galápagos sunshine.

Because the beak of a finch is its can opener, tweezers, and soup spoon, beak size plays a decisive role in determining which seeds a bird can use. *Geospiza fortis*, one of the most variable species, has the beak of Jimmy Durante, and the biggest birds can tackle even the biggest, toughest seeds. Smaller birds of this species take on big seeds too, but their smaller beaks work like hammering a nail with a chopstick—it's worth it only if your life depends on it, and even then it's slow. Although the differences among bird beaks can be measured only in millimeters, the Grants and their students have discovered that small-beaked finches cannot eat the larger seeds as quickly as their big-beaked cousins.

In times of plenty (plenty of rain, plant growth, and seeds), so many seeds of so many sizes and hardnesses abound that most birds can feed themselves well. Subtle differences in seed-eating acumen do not matter. But times of plenty do not last forever. Some years the rains falter and the burning sun beats down on the Galápagos, relentlessly draining the vitality from soil and plants as well as the panting birds. Parched and dormant plants produce no seeds, so the food supply for the population of torrid finches collapses during droughts. Some of the seeds found and rejected in better times begin to look better and better to hungry finches, even the large, tough ones with Fort Knox exteriors. In these skin-and-bone times, small differences in beak size spell the difference between a meager diet and starvation. Birds with the largest beaks live longer, banging out a bare living breaking into seeds difficult to conquer, and the small-beaked birds die first.

Droughts come and go on the Galápagos, leaving behind an

evolutionary legacy written in the hungry brown survivors. These birds pass on the genes for larger beaks to their offspring; as a result, evolution occurs as quickly as the length of a dry spell.

Why Fast Evolution Matters

A light dusting of snow falls as I write this, preventing the residents of Boston from having any thought that March should escort the arrival of spring. But the cold, shallow blanket calms the mind and pleases the eye with its white, granular purity. It lays an unhurried coat over the skid marks of the day.

But when a flurry turns into a fury of blowing snow, calm freezes in the driving storm. The white deluge layers quickly and thickly, indiscriminately covering sidewalks with a slippery icing. Freeway arteries clog, the snowplows rumble their arrival, and a faint survivalist mentality quickens.

Like the rate of a snowfall, the rate of evolution matters too, and not only because evolution's fast pace makes it possible to observe it, measure it, and experiment on it. The rate of evolution is also important because at the faster rate, evolution can happen within our lifetimes. At this pace, we humans can alter the evolution of the species around us, usually making them better competitors, pests, or parasites—at our own expense.

The evolutionary story can be illustrated with guppies and Darwin's finches, or pinned to the tabletop and dissected with experiments on flies and mice. But the real story of evolution has teeth that are visible in the new abilities of species around us, whose manner of making a living has been fundamentally altered by the human world. From microbes that eat antibiotics and viruses that win in the immune wars to insect pests that evade a billion-dollar pesticide campaign, evolution marches onward despite the best intentions of human industry.

The rapid change of common diseases and chemically defended plant pests has been cataloged in great detail. But the day-to-day impact of these changes has been little noted, and the cause—*evolution*—generally ignored. Partly it may be a political and cultural fashion to ascribe these changes to some force other than evolution. Many medical journals assert that the ability of bacteria to kill even in the presence of massive antibiotic doses flows from the *development* of antibiotic resistance. But evolution is the root cause, and the failure to make this technical distinction could be fatal to some poor patient. Why? Because if something develops, then it *just happens*, and we may never know why. In *The Adventures of Huckleberry Finn*, Huck and Jim wonder about the stars—"was they made, or only just happened?"—and Huck opts for "happened" because he can't understand the way the stars might have come to be. And if antibiotic resistance just happens, then we have no notion of *how* it comes to be, and no real chance to block the rise of some of the world's deadliest forms of life. But if something evolves, then the science of evolution can chart the answer to why, and perhaps prevent or change it.

Fast evolution sees to it that species change between yesterday and tomorrow. Species can evolve as their habitats change and they can track advances in human industry, taking advantage of resources we inadvertently provide and parrying the thrust of our attempts to control them. In past decades, the human impact on the ecological world has drawn much deserved attention. Ignored through most of this critical debate has been that humans also have an evolutionary impact and this impact will not be put off like some distant legacy in an uncertain future. Instead, the legacy is today's child with an ear infection or today's stockholders pouring money into a new bioengineering project. Evolution matters in these cases because it happens as quickly as tomorrow, steaming along as a biological chain reaction that we have fueled and that we need to understand.

Chapter Three

The Engine of Evolution

The reception after a student violin recital is a stressful time for us working parents. We come for the music—and to compliment one another's children—but it's really our domestic accomplishments that are up for terrible scrutiny. Across the table of baked goods scatter individually contributed plates of cookies that gossip loudly about our dedication as parents. Ungrateful confections, the cookies turn into loudmouthed domestic spies that reveal scandals about the sort of home we run, because everyone immediately knows which cookies are home-baked and which are not. The store-bought ones have the regularity of hubcaps or any other products extruded from machines—a dead giveaway in the potluck cookie game.

By contrast, good old home-baked cookies are enormously irregular. Round or oblong, flat or globular, the variations can be subtle or strong, but careful observation can tell you that here sits a pile of cookies that came from a real biological source, not an automated factory owned by a tobacco conglomerate. And everybody eats the homemade cookies first.

Variability is a hallmark of biological populations, the real-life reflection of a production line independent of machines. Populations of manufactured goods might pour out of the controlled crucible of a steel foundry, perhaps pounded into molds of unforgiving iron by humorless hammers, but organisms form in the slosh of protoplasm and the vagaries of tomorrow's weather. Like homemade cookies, real organisms in natural populations sport badges of individuality that make them intricately different from one another. The differences can be subtle, like a finch with a thicker beak, or profound, like a whale with legs. But whatever they consist of, these differences add a pervasive texture to the background of the biological world, a texture missing among the manufactured one.

Variation serves as an integral part of the fundamental mechanisms of evolution. Evolution acts as a deep-thrumming engine of biological change and, like any engine, it has fuel, an energy converter and a gearbox that exerts force. Just as your car engine can't let its pistons go on vacation, all the parts of the engine of evolution must function together. If we consider each part separately (like looking at a plastic model spaceship before it's assembled), then we can build a better understanding of each component and not glue the warp drive in upside down.

THE FUEL OF evolution consists of the natural variation between individuals in a population. This variation struck both Charles Darwin and Alfred Wallace, the first inventors of the idea of evolution by natural selection, and became one of the core observations that demanded the birth of the evolutionary idea. Variation remains a prominent feature of populations as disparate as viruses and velociraptors. Wherever you look in the world of natural populations, organisms differ from one another. Thus, the fuel of evolution surrounds us—sometimes piled high like a woodstove's winter diet, sometimes at low ebb like a graduating senior's attention span in class. Like any engine without fuel, evolution without variation has nothing to work on—it comes to a whimpering halt. But the engine's operation demands more than just adequate fuel.

An energy converter is required, a role filled by natural selection itself, operating when the spectrum of slight physical differences alters the lottery of life and changes how an organism lives or dies. Do these physical differences make it more difficult to feed offspring? Do they change how well a male can attract a mate? If so, then natural selection begins to act—altering the makeup of the current generation or altering which individuals have the most young in the next generation. This selection acts like a piston converting the exploding energy of gasoline to the smooth ride of a luxury car: it creates a critical link between the variation of a population and the real-life consequences of that variation.

Inheritance of variation appears as the third element—the gearbox that transmits changes made by natural selection down through the generations so that new generations possess a novel genetic legacy. Without inheritance, the evolutionary engine races without effect, each generation dawning the same as the one before. But when offspring inherit the variations that were successful in their parents, then the next generation can be more successful than the one before.

Evolution by natural selection thus relies upon three interlocking elements that must function together like the legs of a tripod—variation, differences in reproduction, and inheritance. When all three elements appear together, evolution becomes the expected outcome of biological life. Many biological systems shift, change, and evolve as we watch in the lab or in the field. Even replicating molecules, jostling one another in glittering test tubes, can be chemical mimics of the evolutionary process. As long as they possess all three evolutionary elements, populations of simple nucleic acid molecules can evolve overnight into tighter, leaner versions of themselves.

Examples of evolution in the test tube, however, carry less of the thrill of the biological oddities that natural evolution has produced—a seahorse that mimics a floating seaweed just swam through my imagination—and they don't carry with them the deep sense of connection between all living things that is one of the prime dividends of evolutionary thought. But they do show the breadth of power of the three evolutionary principles. They show how the engine of evolution can operate even when the evolving units themselves aren't even alive. And they show that evolution can sometimes fail. Because unlike the person at a party telling you a joke you've already heard, evolution can be stopped—if the fuel disappears, the converter fails, or the gearbox breaks. By understanding the three interlocking elements of the engine of evolution, we can peer more deeply into the limits of the engine's function and perceive more clearly the scope of the evolutionary change we create around us.

The Fuel: Variation Between Individuals

In the gray dawn of a flat day in Glacier Bay, Alaska, while the mists curl and awaken slowly above the water, deep swirls and breaking bubbles tell an experienced boater to watch and wait. As the tide changes its mind, the marine swirls tighten and compress. Abruptly, the inlet gathers itself and then erupts with a fountainous explosion—an explosion of whale. Humpback whales come here to play with their food and nurse their van-sized newborns.

They arrive as regularly as the tourist campers at the nearby Gustavus Inn, and most are known by sight. These whales sport dramatic white and black swirls and blotchings on their tails, as if whale tails were the palette of a doodling Poseidon. So striking are the individual variants that even we dull air breathers can see how whales differ.

These melted checkerboard patterns allowed Scott Baker, then a graduate student at the University of Hawaii shivering in the field-work of Alaskan summers, to tell humpback whales from one another, naming them and keeping books on the movements of each during their yearly migrations from Hawaii to Alaska. To the whales, tail markings may be as meaningless as an ambassador's compliments, but they told Scott how the whales migrated. They told him that the same whales playing in the winter breeding grounds of the turquoise tropics, where males sing soulful songs and perform lumbering acrobatics to entice female approval, can also be seen during the summer in the businesslike gray north where food lies in the water like an all-you-can eat buffet and whales compete to make the most fat. Whales do not leave for-warding addresses and can't be followed from one place to another. But Scott tracked them down, all because every whale tail differs.

Little Differences That Matter

The variation in a whale's tail probably matters very little, but sometimes such physical differences can play a pivotal role in the roulette wheel of life. I happen to be sitting in an uncomfortable seat on a crowded flight as I write this, so my mind naturally drifts to the importance of having the right leg length.

Take the back legs of a small green *Anolis* lizard. The islands of the Caribbean, from the shaggy summits of Jamaica's Blue Mountains to the coral sand donuts that scatter across the shallow Bahamian shelf, support a fleet of agile, tree-living lizards that have also cheerfully invaded most of the southern United States and Hawaii with their quick scramblings. Living among the thin twigs of tree tops and the stout limbs of lower branches, individuals of the same species differ widely in their leg length. Some have longs legs that splay out from their body like a sumo wrestler's stance, while others have shorter legs that more compactly wrap around a narrow branch. Long legs provide greater speed on wide branches, whereas shorter, wraparound limbs confer greater agility in clambering and prowling for insects among the twigs. Given their freedom to choose, lizards will tend to occupy branches that "fit" their legs well.

These slight differences in leg length have important consequences for lizards; habitats with short vegetation and thin branches provide better homes for animals with shorter legs, whereas taller vegetation suits a long-legged lizard better. To underscore the importance of leg morphology in evolution, Washington University herpetologist Jonathon Losos revisited an old experiment in lizard ecology, which had been conducted decades previously on the tiny sand islands of the Bahamas. Caribbean ecologist Tom Schoener had surveyed islands until finding a set that were lizardless, then had added five to ten pioneering animals to each. All the pilgrims had come from the same source population on Staniel Cay. By 1991, ten to fifteen lizard generations after the introduction, Losos and Schoener teamed up to return to the same islands and measured the height of the vegetation and the lengths of the lizards' legs. They found lizards with shorter legs on islands with shorter vegetation. Those where the vegetation was taller had evolved longer legs. Such rapid evolution, fairly common in introduced populations, generated lizards that had the legs for better survival on their new islet homes.

BEYOND LEGS, BENEATH the feathers, scales, skin, or the tweed jacket of your favorite animal, many other types of traits—besides visible ones—vary importantly from individual to individual. Our bodies are bags of chemicals, bottled up in skintight packages, and these chemicals are the products of genes that lie like hidden decoder rings in the Cracker Jack box of our cells. Because we have different genes than our friends, and different genes than our children, our genes produce different chemicals that make each of us a unique soup. The letters in the alphabet soup of the cell spell different words in all of us.

Human blood is a key reservoir of this uniqueness, and humans have a wide variety of blood types. The major way of classifying blood—by blood group—involves a simple lettering system of A, B, AB, or O. Extensive data on such variation has been compiled because blood type was a surprising and hidden killer during early experiments in blood transfusion. Type A blood can not be transfused into type B people without a nasty immune system crisis because the immune system of type B individuals doesn't recognize A blood cells as "natural."

To understand this unforeseen blood reaction, intense work has gone into elucidating the genetic basis of blood type differences. These studies have revealed that a particular protein places key sugars onto the outside of red blood cells. Whether an individual has A or B type blood, both types for those of us that inherited both from our parents, or O type if neither sugar is present, depends on the kind of protein the individual has and how the protein creates the sugar signature of the red blood cell. In turn the protein type is determined by the genes that a person inherits from

Mom and Dad. And so from genes to proteins to sugar-coated cells, individuals differ in a chain of molecular causation that results in a difference in ability to give and receive blood.

BEHAVIORS ARE NOTORIOUSLY fickle and variable. It comes as no surprise that a busload of fifth graders on a field trip all behave differently, or that a good and attentive classroom can turn into a circus when a substitute teacher appears. These behavioral variations have little to do with genetics (although some substitute teachers probably wish they had something to do with selection). But some behaviors are instinctive and hardwired, like a mother bird's response to food-begging from a chick, and these can be inherited by offspring. In these cases, different individuals can have intrinsically different behavior patterns—differences that depend on the genes that an individual has inherited.

One of my favorite children's books, *Horton Hatches an Egg*, rhymes the tale of elephant Horton, who was left to hatch the egg of Mazy the lazy bird while she goes on vacation. Horton ludicrously dedicates himself to hatching this egg, despite its not being his own, even though every two-year-old can see he's better off without the job. By contrast, Mazy has a holiday, vowing never to return.

Mazy's behavior, it turns out, represents a common strategy among birds, which often engage in a selfish and nasty habit called egg dumping. Especially among some species like the cuckoo, a female will lay an egg in another bird's nest, then fly off to let the unknown nest owners raise her offspring for her. When the foreign egg hatches, the foster parents are tricked into feeding the new chick, with occasionally disastrous consequences. Sometimes, their own chicks starve as they stuff food down the maw of an alien nestling much larger than their own.

To help save themselves from being so gullible, birds in areas where egg dumpers (sometimes called nest parasites) commonly occur have evolved inspection behavior for suspicious eggs. Spotted cuckoos in European forests frequently lay eggs in nests of the common magpie. Magpies can't count eggs when they come back to their nest, and therefore can't detect when a cuckoo's egg has been added. Instead, they inspect all eggs in their nests and heave any that look wrong over the side. Magpie parents, by being picky about the types of eggs they permit in their nests, save themselves the risk of raising an expensive foster offspring.

But magpies pay a cost for this pickiness. Eggs of different birds can look very similar, and nest parasites often have evolved egg coloration patterns very close to those of the birds whose nests they use. This evolutionary convergence makes it difficult to tell a dumped egg from your own, leading to many mistakes. Imagine coming back to the nest and discovering your mate has heaved out the egg you just laid, thinking it just a bit too different in shape or color. This takes the wind out of the sails of reproductive success, and birds that engage in egg heaving only have an advantage over nonheavers in forest habitats where nest parasites live. Where there are no nest parasites, nonheavers will have more young. Studies of magpie behavior in Europe, for example, have shown that where nest parasites like cuckoos abound, egg inspection dominates the magpie's behavioral repertoire. But where parasites are rare, the local birds do not inspect and heave eggs. These behaviors seem to be genetically based—somehow dependent on an inherited tendency to inspect and judge egg quality—and changes from forest to forest represent the evolution of egg-heaving behavior.

In fact, we have seen how such variation allows behavior to evolve. Small birds called village weavers show strong egg-inspection behaviors in their native Africa, where nest parasites abound. In the seventeenth century, they were introduced to the Caribbean island of Hispaniola, where no egg-dumping nest parasites exist naturally. The transplanted weavers quickly lost their egg-inspection behaviors, and were then an easy target when the egg-dumping shiny cowbird invaded Hispaniola in 1972. In the 1980s these nest parasites were shown to have a terrible effect on weaver nest success. But by 1999, the weavers had re-evolved their egg-inspection and parasite-heaving behaviors, rejecting 89 percent of cowbird eggs and reducing the impact of cowbird parasites on weaver reproduction.

In magpies and weavers, abundant genetic variation, just like the other types of variation, can be a fuel for evolution. But variation by itself won't suffice—the differences have to matter, leading to differences in how well individuals reproduce. These differences bring us to the pistons of the evolutionary engine—natural selection.

The Energy Converter: Natural Selection

I am no longer allowed to do laundry. Too many shrunk and ruined shirts of mysterious and fragile fabrics have been delivered from my hands, too many pinks live where white used to reign. I am inclined to accept these laundry failures as an acceptable form of natural selection, one that Mother Nature and Macy's meant us to participate in. But other people, it turns out, don't see it this way. Only some of us possess the dispassionate ability to see why a whole class of fabrics must be culled from the wash. We are the laundry impaired. To others, every sock is sacred, whether it has a match or not.

To me, something about laundry suggests a live-and-let-die philosophy that says all your clothing should be equally hardy vis-à-vis the simplest wash and dry procedures. Hot water, cold water, laundry soap caked by age . . . some clothes can be washed just about any way, including smacking them on a rock in a stream, without damaging them. These fabrics are the hardy ones, and when you possess only the toughest clothing, very little attention need go into the conditions of the spin cycle or the temperature setting on the dryer. It's a simple—if somewhat rumpled—style, even if a few shirts must die to preserve it.

But if you buy a shirt that requires the Houston Space Center to get clean, then after the first wearing, you toss it in the wash and pay for your mistake when it emerges a color and shape Salvador Dalí couldn't imagine. A tragedy of course, but the loss serves a useful purpose because along the way a weak and frail article of clothing gets culled from the healthy, roaring pack of your regular stuff and your laundry returns to its original robust state. This represents selection—a special kind of selection not envisioned by Darwin, who never once did his own shirts. I call it laundry selection, although my family calls it idiocy.

NATURAL SELECTION CAN be defined as the differential survival or the differential reproduction of individuals that are physically, physiologically, or behaviorally different. An individual can reproduce more if it lives longer, or it can reproduce more by having more surviving offspring in each breeding season. As a result, selection can occur when a characteristic of an individual changes how long it lives or how many offspring it can have in a season. Because biological individuals often vary from one another in important ways, opportunities for such selection occur very frequently.

I've already discussed the first critical element of this idea—the common variation in all manner of organismal features among individuals within species. We can easily observe the second element—that these variations make a difference in the success of a particular individual—in many natural and human-made settings. Selection due to poor food supply, outrageous mate-choice preferences, or other features of the daily lives of animals, plants, fungi, bacteria, and politicians all can create the circumstances that allow some individuals to reproduce more than others. Examples of natural selection in action can be sifted from populations of fish, beetles, annual plants, and bacteria. Selection acts on mate-choice traits in lizards, guppies, birds, sea urchins, and plants. Human technology is the prime generator of selection for viruses, bioengineered crop plants, families of commensal rats, and the invading weeds of a thousand roadsides. This book documents the facts of many such cases, but to understand selection we need to see where its power comes from and how it can cause such change. Our first look will be into the eyes of age.

Why an Elephant Never Forgets

We take aging for granted, being used to its effects on the people and pets around us. But in the natural world, we almost always see youngsters or animals in the middle of their reproductive years. Even if by stroke of rare fortune or a PBS special, we should see a senior gorilla salted with a silver-backed pelt, this gray badge signifies a male at the prime of life and height of social command, not the declining years of a gorilla grandpop. The birds on the wire overhead seldom need a cane, and arthritis is not a common complaint among kangaroos.

Where are the aged, postreproductive, grandmother crows? Elephants beginning to forget things? Sharks losing their teeth? We don't see them often—unless they're our pets. Yet most animals can live into their "golden" years if we protect them. In the classic *The Biology of Aging*, Alex Comfort details that animals can usually live longer in captivity than they do in the wild. A horse lived to sixty-two years of age in Latchford, Great Britain, in the nineteenth century, and a male orangutan lived to be fifty-three in the Philadelphia zoo. *The Guinness Book of World Records* reports the record for dogs and cats as twenty-nine and thirty-six years, respectively. Goldfish have kept their owners company for five decades, if "company" is really the right word.

Lions in zoos may live comfortably for many decades, but studies of lions in the wild show none old enough to collect Social Security. Females live longer than males, but seldom longer than twenty-five years. Moreover, neither males nor females tend to live beyond the age of reproduction. Sulfur-crested cockatoos can live eighty years in captivity, but the raucous cries you might hear in an Australian forest today come from mere teenager cockatoos, not octogenarians.

If animals can live far longer than they generally do, what keeps them from doing so? The simple answer is that they die young. In the wild, death rides shotgun beside predators, parasites, environmental extremes, lack of food, and a cadre of other callous conditions, all preventing animals from living as long as they might. Records from small-bird studies describe the lottery of a wren's life: about half of all birds do not survive their first year to breed in the second. Of the half that do return, only half will return the following year. Few of these birds live to be more than four years of age—none worry about retirement.

THE SIMPLE FACT of death by a thousand dangers augments the power of natural selection, because death often strikes nonrandomly. The fallen sparrows of Hermon Bumpus's 1898 study of natural selection show that New England sleet storms wreak havoc on unprotected flocks, delivering death to some sparrows more easily than others. Predators also exert strong choice at meal times. Thrushes spy out and consume garden snails most successfully when the nervous escargots show particularly vivid colors on their shells. The snail shells that a thrush can easily see, striped and colorful like a child's racing car, quickly become shattered statistics, proving the danger of being seen. More drab, the intact shells of the survivors show few brilliant splashes—these dull snails slurp their vegetable dinners like gulag prisoners and maybe even dream in olive and brown. In these cases, death discriminates because it removes some individuals from the population of breeding adults on the basis of those individual's traits.

The Race to Reproduce

Selection does not come only on the heels of death. It is also written in the catalog of life, because sometimes selection is not a struggle to survive, but a race to reproduce. How to run this race can vary tremendously, and this kind of variation is also the fuel for intense natural selection.

Along the inner Coast Range of California, the small, side-blotched lizard makes its cold-blooded living by waiting for winter's retreat each year. In the spring, females lay their eggs, large globes so packed with yolk that each one represents a serious investment for the young mother. Females lay from two to seven eggs at a time, produced in monthly clutches throughout the summer. Since a dedicated mother uses up to a third of her body weight to produce her eggs, she might lay her own weight in eggs during a summer.

Females vary in the size of the eggs they lay, but they also vary in egg number. Here lies an example of one of biology's great compromises: using a limited amount of food energy to make eggs, a female can make many small eggs or fewer large ones, but she can not usually make many large ones.

Given this trade-off, how many eggs should a female lizard lay? Reproductive output is higher if she makes many eggs, but this is not the only consideration. Larger eggs make for larger offspring and bigger babies run faster and often have a higher chance of survival. If the survival of young is poor, producing fewer, larger offspring may be a better reproductive strategy than spewing out a fleet of subcompact babies that have a higher chance of death.

Barry Sinervo at the University of California at Santa Cruz has investigated this reproductive teeter-totter in California lizards for years, showing that natural selection acts swiftly on different reproductive strategies. In years with low rainfall, he and his students found that females do best to produce large young from large eggs. Not only is survival of the females that make large eggs higher, but large offspring survive better. During a drought from 1989 to 1992, average egg size increased about 20 percent. But the California rains came in 1992 and changed the rules. When rainfall was plentiful, selection favored production of smaller, more numerous eggs. Selection was so strong that by 1994, average egg size had declined below the level egg size in 1989. Thus, all the evolutionary change that had been wrought by drought had been undone by rain.

What Alternatives to Selection?

If every individual lived out its full allotment of years—if we all reproduced and died on schedule—like a female octopus does after tending her eggs—then the opportunity for natural selection would be reduced. The opportunity for some individuals to survive longer on the basis of thicker fur in glacial winter, deeper roots in a howling storm, or a color like a mottled twig when an insect-eating bird patrols by can make one organism the winner of life's daily struggle, and leave a different individual out of the pageant of the next generation.

How could selection *not* work? If every life was exactly the same length, if every death was accidental or predestined, if survival had nothing to do with the cards an individual brought to the game, if the grim reaper was blind and death landed at random, then maybe selection wouldn't occur. If every organism had the same chance of having offspring—with no reproductive bonus for parents that could gather more food, fight off predators, or shelter a nest through a storm—then maybe selection would stop. If none of the differences among the individuals of the world make any difference to their legacy of offspring, if fighting your way up the river isn't required to spawn, then maybe selection wouldn't be common, and maybe evolution would stop. However, although many differences between organisms don't change the accounting of offspring—like the uniqueness of a whale's tail—enough matter to keep the evolutionary engine running on and on.

The Third Element: Heritability

If you like Doberman pinschers because of their stubby, cute tails, you might be surprised to see a litter of Doberman puppies wagging theirs. The puppies start out with normal long tails, but will be treated to a small operation to make them cleave to the kennel club view of Doberman dogma. But the dogs' genes win in the end, as generation by generation, they produce puppies with tails just as long as the original Doberman stock. Clipping Doberman tails does not lead to evolution, and if you choose your breeders on the basis of their nice stubs, too bad—the puppies will disappoint you and still need the chop.

IN PLENTY OF other cases, evolution does not result from selection because the characteristic in question isn't passed on to offspring. I have a cheerfully thoughtful friend who in graduate school faced a population explosion of raccoons while living for a time on a small island in Puget Sound in Washington State. He would have generated a whole population of blue raccoons—except for the rules of heritability.

A land steward for a conservation organization, my friend had perched himself on a water-circled hummock of rock and meadow, trying out a pioneer life without electricity or running water—except for a twelve-volt stereo TV and a tank of water in the wood-burning sauna. One night, his pioneer serenity was interrupted by the arrival of a pair of raccoons. The circus stripes of a raccoon hide a stubborn, greedy, and vicious disposition; on a small island, they soon exhaust their food supply and their welcome. No garden, no pantry, no scuttling crab under an intertidal rock sits safe from a pair of hungry raccoons. And the original pair soon turned into a half dozen, still taking cute lessons every day and raiding the garden every night. They were a scourge and were wrecking the little island, but killing them was out of the question—it was a nature preserve. The raccoons had to be trapped and taken to some other island to bother somebody else.

My friend was a graduate student through and through—dedicated to data and simple experiments and testing hypotheses about the world. Here was the chance for a simple experiment with big practical implications. If he trapped the raccoons and delivered them to another island, would they ever return? Could they find the island again and swim back across the cold waters, across the deep currents? Because raccoons don't have individual markings like whales, identifying any of the original residents if they returned to the island would be impossible.

There was only one simple way a good biologist could resolve this dilemma: to paint the raccoons some color—say, blue—and release them on another island. Not entirely blue, because that might hurt them, but blue enough, here and there, to be unmistakable and recognizable should they return.

The result of this impromptu experiment was soon clear: the blue raccoons returned very quickly to their original home, taking up their ecological ravages where they left off, and continuing to make baby raccoons at an unsettling rate. Did these blue raccoons go on to produce litters of blue youngsters and populate the island with blue adults? No, because blue paint is not genetically determined; as soon as it wears off, the blue is gone forever, unable to play a role in the evolutionary trajectory of the rainbow of raccoons in future generations. Even though an island full of blue raccoons was selected, the stylish effect was only temporary. The story ends happily, at least for the raccoons. The gardens never recovered and the last I heard, the raccoons were sole proprietors of the island.

What Gregor Saw

Unlike the blue of these unwelcome raccoons, color in other animals often results from the action of genes that can be passed on to offspring. Positioned deep inside the cellular instruction manual, coloration genes act by conspiring to make pigments and then using them to paint the outside of an organism—its skin cells, hair folli-cles, or feathers—to more or less precise specifications. When these genes pass from parents to offspring, the offspring tend to look like their parents. Therefore, we call color pattern a heritable trait.

In fact some of the first genetics experiments were done to try to understand how color is inherited. Gregor Mendel drew the first secrets of genetics from peas he planted in a monastery in nine-teenth-century Austria. He noticed two stunning features of the plants he tended—features that we all see and take for granted, but that told him the rules for how genetic information is used to build a plant. In the nineteenth century only Mendel read the clues—only he would give peas a chance.

Peas come round or wrinkled, yellow or green, but few peas sport intermediate colors, and few sit midway between round and wrinkled. This was the first telling observation in Mendel's garden. If a green pea plant receives pollen from a yellow pea plant, no greenish yellow peas result—the parental colors do not blend. If they did, then soon the cross-pollinating between colors would mix them all together and produce a yellowish-greenish mush. All peas would be this dull color, like old poi at a tourist luau, and the pastel greens and yellows of the peas in Gregor's garden would never be seen.

Instead, Mendel noticed that yellow and green peas, when crossed, sometimes produced yellow offspring and sometimes both yellow and green offspring. Blending did not occur—pea color seemed to be a choice of several limited options, like what cable TV company to use. This would have been exciting news to Darwin, who thought (like the rest of the contemporary scientific world) that heredity usually produced blends and who worried that the "principle" of blending inheritance was a sharp thorn in the side of arguments for the power of evolution by natural selection.

Mendel's observations on nonblending inheritance were just what Darwin needed, but Mendel went further. He noticed a strange additional fact that helped explain why inheritance was not a blending process. Sometimes a trait disappeared in a generation, only to reappear in the next one. Sometimes a cross between a yellow and green plant produced all yellow offspring. But breed two of these offspring together and sometimes the green would reappear! Mendel reasoned that something that caused the green color was sitting in the yellow plants. Lying dormant like thoughts on Saturday morning, the green trait had been passed from a green parent pea to its yellow offspring. Then it was passed on to the next generation, at which time, occasionally, it began to function again.

Through careful experiments, Mendel worked out the rules for the passage of these elements from one generation to another. He deduced there must be two markers for every trait in every individual and that sometimes one was dominant over the other. But his biggest accomplishment was figuring out that they existed at all, describing them as units passed from one generation to the next. Today, we call these units of information *genes*. Darwin could have profited hugely from this insight, but the results of Mendel's experiments remained unknown for a generation until being rediscovered long after Darwin's death.

One Gene, Two Genes, Three Genes, Four

Of course, Mendel took some of the fun out of heredity—he was a monk, after all. Rather than some mysterious "plasm" that formed during a sexual union, heredity was a mere bookkeeping of genes passed down like heirlooms to offspring. Nowadays, genetics has become much more complicated as our understanding of the molecular basis of genetic action has grown. And even Mendel understood that some traits, like flower color, seemed to be under the more complicated control of two separate genes. But the basis of our understanding of inheritance remains the transfer down the generational hierarchy of genes inherited from past generations.

As our understanding of the complexity of genetics has grown, we have solved the problem that had stymied Darwin and his friends for so long about inheritance. Sometimes so many different genes control a trait that it looks to be inherited in a blending fashion—but even in these cases, genes are passed down intact through the generations.

Typically, many genes act on traits important to natural selection, meaning that each gene usually has a small effect and that the final trait sums up the effects of each of the genes that an individual inherits. Such traits, called quantitative traits, tend to vary continuously, without big jumps in character state like green to yellow. For example, in people, height is a quantitative trait, and we come in many different heights.

How do people get to be a particular height? As a result of sexual reproduction, offspring combine chromosomes from both parents, and each gets two copies of all the many genes that influence height. A tall parent will tend to provide "tall" versions of the information content of these genes. Tall parents tend to have tall children because the children have inherited "tall" gene versions—called alleles—at many of the genes that control height. A union of tall and short parents will often generate intermediate offspring as each parent gives to the next generation a set of "tall" or "short" alleles.

But the inheritance of height does not strictly average the genes from both parents, and like splitting a dinner check ten ways, sometimes things don't add up. Tall children can sprout from short parents (for example, all of the tallest people in the world had parents shorter than they) because genes get packaged randomly into sperm and egg; therefore, some offspring of short parents can inherit a larger than normal share of the few "tall" alleles those parents have. If they also got many "tall" alleles from their other parent, then they can be taller than both. Such events occur often in families showing that inheritance can't be strictly blending.

But the environment can also influence many traits, thereby diluting the similarity of offspring to parents. This dilution comes about for familiar reasons, especially for traits like height, which can be affected very strongly by food supply and good general nutrition. Near my home in Lexington, Massachusetts, there persist proud houses from the era of the Revolutionary War, lovingly restored and persistently watched over by local historical societies. Don't, however, take too fast a tour through John Hancock's old house, because many of you would crack your head on the door lintels. Doors in these old houses were built for smaller people. Have we evolved into giants in Lexington? Probably not, but with a better and more consistent food supply, our bones grow longer and our height increases.

Putting the Engine Together: The Gall of Those Flies

All the preceding has been like a surgeon's Thanksgiving—a detailed dissection of the evolutionary turkey to reveal the separated parts (variation, selection, and inheritance) and how they all look when laid out in full view on a platter. A full understanding of evolution should finish up with a description of the way the elements fit together, providing a view of the whole turkey, not just the pieces.

VISIT A FIELD of goldenrod. Especially in the early fall when the reedlike, upright stalks of the goldenrod plants sway coldly in the first autumn winds, you can see a veritable hotbed of evolution in action. At that time, when frosts gather like children at a story-teller's feet, you will often find odd-shaped, marble-sized swellings visible two thirds of the way up many of the drying stems. Called galls, these globes mark the homes of tiny insect grubs that live by chewing away the center of the goldenrod stem.

Plants make the galls, encapsulating the sucking impropriety of the insect's diet, but the insect still lives comfortably protected in the middle of the globular mass, sucking its sappy food while sur-rounded by an opaque cellulose sphere. There, the grubs spend their childhood, growing large on the plant's bounty, finally trans-forming in the fall into patient pupae waiting out the winter. The following spring, they burrow out of the gall, emerging as adults, to mate and lay eggs for the next generation.

The galls come in different sizes. The plants themselves vary in the sizes of the galls they allow—some clusters of goldenrod pro-duce huge globes compared with small ones elsewhere. But the insects make a difference too, varying in how big a gall they induce. Some grubs cause a plant to make a prodigious gall. Other individuals generate small quarters, living hemmed in by close walls. This variable size of the galls fuels a surprisingly intricate evolutionary dance.

Why They Call It Grub

Despite the grub's sheltered life, encased in a cellulose dome, ene-mies can still attack the insect, and a nasty fate lays in store when a female wasp called a parasitoid interrupts the peaceful existence of a gall. The people who first conceived the movie *Alien* must have known about these wasps, because their babies, like those of their Sigourney Weaver–chasing screen cousins, viciously consume other creatures from within.

A female parasitoid wasp begins her attack by inserting her narrow ovipositor tube through the wall of a gall, laying a single egg inside. The wasp egg hatches into a hungry wasp larva, which blindly finds the gall insect and burrows deep inside it—feeding voraciously until it consumes the grub. The wasp then metamorphoses from larva to pupa to adult, and chews its way out of the gall to emerge the following spring, ready to mate immediately and lay eggs in more grub victims. But the newly emerged adults cannot kill all grubs. The walls of a big gall provide secure protection from wasps, because the thick gall walls are wider than short wasp ovipositors can penetrate.

Going out in a field in the early fall, you can gather up the galls and see the fate of each. A gall grub bores its way partially out of the gall early in the fall, leaving only a thin crust covering the escape tunnel's opening. Wasp larvae kill the grub before this useful tunneling, and so once they are adults they must bore their own way out. Cracking open the galls collected from autumn fields, you can see which ones have gall fly grubs inside and which harbor wasp larvae. This kind of quick field census shows that wasps usually emerge from small galls; insects that make larger galls are protected from wasps and survive the threat almost completely. This difference in predatory ability generates strong natural selection: insects inside large galls are less likely to be killed, whereas small galls disappear from the population.

YET LIFE IS no picnic for larvae in big galls. Although they are protected by thicker walls from threatening ovipositors, and thus safe from wasp attacks, birds often attack larger galls by breaking them open in winter and picking out the grubby contents. Even though by winter the goldenrod plant has died, the dry stems persist, rattling through the snowstorms and winds of winter. Goldenrod galls serve up the tasty grubs of a December lunch for hungry birds, and one by one, the galls are visited like donut shops on an overnight drive. The grubs are pecked out, picked up, and swallowed, as the birds leave the signs of a careful attack in the punctured gall on the swaying goldenrod stem. Especially in winter, woodpeckers use the insect tunnel bored by the grub to neatly extract a meal. Because the birds focus on larger galls, this kind of predation also exerts natural selection.

Between wasps and birds, a large proportion of the gall insects may be eaten, with 50 to 90 percent of the galls being attacked in any single year. This rate of mortality, especially when so tightly linked to variation in the galls, provides a consistent and strong opportunity for natural selection to operate. But the actual outcome of this selection depends very strongly on which predators are abundant in any one year. In a year with few wasps, the average size of the surviving galls will decrease during winter, when birds ravage the old fields looking for Christmas dinner. In a year with few birds, wasps will cause an upswing in average gall size. In some years and places, there might be a balance between the two sources of natural selection so that, despite intense size-selective natural selection, no real change occurs in the population. Gall size as an evolutionary characteristic sits on a teeter-totter, shifting back and forth depending on the weight of selection on either side and on the other organisms found interacting together.

From this example, we can learn three things of value. First, mechanisms of selection—in this case, how size-selective predation occurs—almost always involve detailed features of the natural history (or lifestyles) of the species of interest. Careful observation produces this kind of information, which usually requires an understanding of many facets of the organism's ecology. Second, mortality based on size is strong enough to be easily observed in this system. Similarly strong bouts of natural selection have been recorded in many other settings, suggesting that this example represents a large number of cases in which natural selection acts over short time frames. Third, natural selection is exerted on this system from multiple directions. There is not just one selective agent operating, but two. Given that most organisms live in a biologically rich and environmentally complex world, natural selection has many potential sources, many that might be acting at the same time.

Evolution Across the Generations

Changes in gall size herald the bold signature of natural selection in the goldenrod field. But it still takes the third element to make evolution happen: there must be inheritance of gall size from the end of one season to the beginning of the next. The plants somewhat determine the size of the galls on their stems, but like a careful editor at a high-class publisher, the insects also have an element of control. In this case, grubs that make a large gall develop into adults that lay eggs that in turn hatch out into offspring which also tend to force a plant to produce large galls. Likewise, a small-gall individual tends to have offspring that in the next generation also tend to induce goldenrod plants to make small galls. In this way, selection during a season will result the next year in changes in the population of gall insects.

The continuity of a change, effected by natural selection and transmitted across generational lines by the accounting of genetics, allows the steps of evolution to be laid down one by one. This was

Darwin's insight—that evolution stepped along on the three coordinated legs of variation, selection, and heredity. It was also Alfred Wallace's insight, gleaned on a feverish night in 1858 in the jungles of New Guinea. And it was also the insight of the virtually unknown Edward Blyth, who published an article in 1837 about the role of natural selection in evolution. But Blyth focused on natural selection's ability to weed out unfavorable or extreme variants, leaving behind only well-adapted individuals. Hermon Bumpus's sparrows, blown to earth by that New England winter storm in 1898, felt the wrath of this type of selection, as birds with odd morphology succumbed.

Darwin's view of selection was different; he perceived it as a generator of new biological adaptation. He conceived a great—and unintended—evolutionary engine that hummed, vibrated, and occasionally sparked, and ran throughout the corridors of time. It ran for every species all the time, sometimes quickly, sometimes slowly, sometimes yielding up trivial or invisible changes, sometimes delivering a cargo of startling innovations. It ran forward and backward, increasing body size and decreasing it, filling the land with dinosaurs, then pushing them aside. It ran up against limits of body size, running speed, ability to keep warm in a polar winter. It stumbled onto schemes of great efficiency—like the heat-saving countercurrent circulatory system of a tuna—and imposed schemes of enormous waste—like the vast carbon stores in tree trunks all accumulated for reasons of selfish competition for light. It led to make-do solutions and unimaginable innovations, from the aerodynamic disasters called puffins to the extrasensory ability of dolphin sonar.

All of this follows from the three principles we've covered in this chapter: variation, selection, and the tendency of offspring to be like their parents. With limits and unexpected complications, with the coin-toss effects of chance thrown in, evolution by natural selection pervades biological systems. It's built into the way we earthlings—all of us—are made, and it's difficult to stop entirely. Especially when we impose so much mortality on the biology around us, we heighten natural selection and force it to take steps to thwart us. How this happens seldom surprises us. That it happens so quickly can surprise those who think evolution only takes small slow steps. But the evolutionary engine that sculpted the ancient reefs of fluted clams and sharpened the fangs of the saber-toothed tiger also walks the halls of our hospitals and promotes every new generation of crop-eating bug. Now that we know how the engine works, let's see what it can do.

Chapter Four

Temporary Miracles:
The Evolution of Antibiotic
Resistance

If you're a parent, you've acted out a common scene with an unhappy, crying baby at the end of a feverish evening: The world has constricted to a narrow migration between the crib and rocking chair. You walk a parent's metronome, back and forth and back and forth, ticking away the width of the night. In the dim yellow night light, even the Care Bear blanket looks worried as the minutes turn to fussy hours. Your emotions flip channels from concern to fatigue to irritation to worry as you look at the phone, willing an impossible midnight call from the doctor. Stand and walk and pat and hum. This is definitely getting to you. It's another chest cold.

The next morning dawns with the same yellow color as the night light. You've finally gotten through to the family pediatrician, and you schedule a comforting session with her. But once you get there and you're about to gratefully turn the problem over to somebody else—somebody with a prescription pad in her hand—you get the bad news. This time there will be no drugs.

"Probably viral. It's just a cold. Baby's big enough now—he'll clear it on his own," she reassures you.

Stunned silence. Then: "What about antibiotics?" Up to now, the baby's been nursed on an ear-infection, chest-congestion diet of amoxicillin and you expect drugs, you *need* drugs to make your baby healthy and your nights sleepable.

"He'll clear it on his own, and you don't want to use antibiotics for viral infections," the pediatrician answers. "If he's not better in two or three days, come back and we'll reevaluate him."

"Two or three days? Can't you just give him some amoxicillin now? Just in case? Listen to him!" A little panic creeps back in.

"No." The pediatrician is firm. "Because if he doesn't need the antibiotic, he's better off without them—they can lead to drug resistance."

Resistance. A new monster to haunt the parental lexicon, learned later than jaundice, SIDS, and don't-face-the-car-seat-forward. But what does it actually mean? And why does it change the medicines you use?

A year and a dozen colds later you finally get the idea. Resistance of bacteria to the drugs can happen *in your kid*. And once it does, the antibiotic isn't "anti" anymore—the bacteria evolved, because you forced them to.

True, other antibiotics exist, drugs that the bugs can't tolerate yet. You could always use those instead, setting the resistance clock back. But some of these drugs cause more problems—some cause allergic reactions, some kids throw up, and others have diarrhea for ten days until the end of the dose. Not a pretty picture, and not worth the unnecessary luxury of antibiotics on demand.

IN SMALL CHILDREN, viruses, not bacteria, cause most chest and head colds. Because antibiotics combat bacterial and not viral infections, a good pediatrician will not always prescribe these drugs for colds. Instead, you need to wait for them to clear naturally. This is modern medicine in action. The weapon of choice is a power you dare not capriciously wield, because a mild dose of the same drug, used intermittently for a long period, creates the natural selection pressure for bacteria to evolve a defense. By using drugs when the body's natural defenses suffice, you accelerate the evolution of resistance. You create a disease organism more difficult—and sometimes impossible—to deal with.

"That's Funny"

The Fifty-first Psalm sings of the herb hyssop, a veritable medicine chest with leaves so curative that ancient remedies described laying them on open wounds to help healing. Today we know that the herb harbors a fungus that produces a natural penicillin, so perhaps this ancient remedy was the first human use of antibiotic therapy.

Louis Pasteur was thought to have stumbled across antibiotic activity in his exploration of the tiny bacterial cells that he linked to disease, but the first of the major antibiotics, penicillin, was officially discovered by French medical student Ernest Duchesne in 1898. But the medical world did not recognize the value of antibiotics, and thirty years elapsed before the time was ripe.

In 1928, Alexander Fleming grew bacteria in a lab at St. Mary's Hospital in London. Bench tops were covered with bacterial cultures growing as round yellow buttons spreading over flat glass dishes. The cells were of a wide variety, cultured from everywhere, collected like a pigeon fancier collects rare breeds of homing pigeons, including the human pathogen *Staphylococcus*. Few sloppy bacteriologists survive for long—nature selects against them—but Fleming smoked like a wet fire and kept his laboratory door open. Downstairs, mycologist C. J. La Touche had isolated a strain of the mold *Penicillum notatum,* growing it in profusion, and mold spores wafted out of La Touche's lab, up the stairs and into Fleming's open door. The mold, a common stale-bread variety, grew well on Fleming's bacterial plates and soon clashed with the colonies of bacteria already in residence there. No well integrated community of stuffy *Staphylococci* and lowly mold was to be— instead, a microbe war began.

Returning from a holiday in September 1928, Fleming cleaned up the old bacterial plates, including the set of disappointing cultures that had been contaminated by La Touche's migrating mold. Resigned to cleaning up the mess, Fleming dumped the plates in a vat of lysol to kill the cultures, but then a former protégé, D. M. Pryce, appeared at the lab to see how the experiments were going. Fleming fished the cultures out to complain about contamination, and, while showing them to Pryce, finally noticed the odd halo around some of the mold.

"That's funny," he mumbled, puzzled. In a thin band around the *Penicillin* colonies, he could see an unexpectedly clear zone where no bacteria survived, a no-man's-land laced with poisoned broth that killed bacteria. Where the mold grew, the halo preceded it like army scouts before a battle, destroying bacterial cells and clearing room for the mold invasion.

Fleming was intrigued. Nothing interests a farmer like the reasons a crop won't grow, and Fleming set out to discover the cause of the bacterial halo. But not all bacteria were affected. Gram-positive bacteria were killed, but gram-negative bacteria were not. The difference between these two types of bacteria resides in their cell walls, so something in the mold acted on some cell walls but not others. Fleming fussed with this result, but eventually abandoned it when it seemed to him that mold extract could attack only bacteria growing on surfaces. Not until a decade passed were Fleming's extracts shown to kill bacteria growing inside infected mice. Attention gathered quickly to these results and the search was intensified for the agent that killed bacteria.

Fleming's mold extracts held powerful antibacterial activity, in the form of the chemical we now call penicillin, but it was very difficult to purify the drug on a large scale. Finally, a chemical modification of the extract generated a slightly different version, penicillin G, that was easily and cheaply produced on an industrial scale. These discoveries, from a team led by E. B. Chain and H. W. Florey, gave birth to the first of the modern antibiotics. But wartime England was no place for the massive production of penicillin, and the precious mold cultures were spirited to the United States by Florey and his colleague Norman Heatley in July 1941 via the neutral port of Lisbon. Worried that the brown powders would fall into the wrong hands, and afraid to call attention to themselves by carefully transporting glass vials of mold culture on a transatlantic flight, legend has it that Florey and Heatley doused their clothing with the mold—where it no doubt joined the normal inhabitants of tweed—and flew to deliver their cargo. However it got to America, by 1943, penicillin production was in full swing, and the antibiotic era was born.

PENICILLIN REVOLUTIONIZED TREATMENT of infectious diseases and launched a scientific revolution. Many infections, especially the devastating sepses of war wounds, could be rolled back for the first time. Serious and debilitating diseases, previously incurable, now faded away within days or weeks of treatment, saving millions of lives and preventing a great deal of suffering in millions more.

Even some of the world's most heartbreaking childhood diseases vanished under the antibiotic wave. Scarlet fever, caused by

Streptococcus bacteria called Strep A, usually starts in the throat of a small child, and has long been identified by an accompanying red rash. When untreated, scarlet fever triggers the body to develop a destructive immune reaction, damaging heart tissue and bringing on acute rheumatic fever. But the dread childhood nemesis succumbed to a dose of 10,000 units of penicillin a day, together with companion antibiotics like erythromycin. By the 1970s, Strep A had cried uncle and scarlet fever had dropped off the medical radar screen.

Penicillin and the growing family of antibiotic drugs were quickly recruited to treat other debilitating diseases like gonorrhea. By 1953, a decade after widespread use of penicillin began, the stable of antibiotics included chloramphenicol, neomycin, tetracyclines, and the class of drugs known as cephalosporins. Better yet, these antibiotics functioned in different ways to strangle bacterial growth and were effective against a different spectrum of infectious agents. The antibiotic era was in full swing.

And so these gifts—miracle medicines—were used when needed, and used when not. The development of antibiotics followed logically and perhaps inevitably from Pasteur's original discovery that bacteria caused many diseases. Antibiotics were the ammunition; and the war against infection was on. But in this war, which seemed so one-sided at the time, bacteria had evolution on their side.

Fly in the Ointment

Penicillin works by preventing certain bacteria from building well-integrated cell walls. Bacteria need cell walls to maintain their shape and integrity, but bacteria trying to grow in the presence of penicillin puff up and die. Nevertheless, by the late 1940s, some bacteria were not puffing up and dying—they resisted penicillin's effects, living through an antibiotic dose that would have killed their ancestors. The common human gut bacterium *Escherichia coli* quickly developed resistance. So did shigella, one of the bacteria causing dysentery, as well as the bacteria causing pneumonia and gonorrhea. In 1991, a virulent and aggressive strain of Strep A began showing up in hospitals. Instead of 10,000 units of penicillin a day, treatment of some Strep A cases now demands 24 million units; even then, the infection can prove fatal, growing so quickly that penicillin does not halt it. Other human diseases treated with other antibiotics have shown similar trajectories. The bacterium causing leprosy, *Mycobacterium leprae*, has been treated for decades with dapsone, but a resistant strain appeared in 1977 in Ethiopia; by 1987, up to 40 percent of cases in India, France, and China were resistant.

Almost all disease bacteria have evolved some level of resistance. A major cause of postoperative infections, *Staphylococcus aureus* has always been a critical complication in surgery patients. In 1952, virtually all staph infections yielded to penicillin, but this percentage fell quickly. Luckily, the antibiotic methicillin uses a different strategy to prevent bacterial growth, so by the late 1960s many hospitals switched from penicillin to methicillin in treating staph infections. Unfortunately, a methicillin-resistant staph had already evolved in Cairo in 1961 and the 1980s saw a large number of infections by methicillin-resistant strains. Of about 20 million surgeries performed in the United States in 1992, about 90,000 cases of staph infections developed, most of them resistant to methicillin. In fact, some bacterial cells had acquired multiple-drug resistance, withstanding most antibiotics commonly used in hospital settings. By the mid-1990s only the advanced antibiotic vancomycin was reliable enough to treat all staph infections, and in 1996 partial resistance to this drug was finally reported.

High-grade resistance can change lives. A few years ago, my wife had an elderly patient, call him Mr. Lee. He came in for lung surgery and stayed in the hospital for the last three years of his life. He'd developed a staph infection resistant to everything but vancomycin, but even the drug of last resort wasn't enough. No nursing home would take him for recovery, and he couldn't go back home—not with a persistent infection that required IV antibiotics. To prevent spread of his high-risk infection to family and friends, Mr. Lee waited out his final years in his hospital room, visited only by Mrs. Lee, enveloped in a disposal hospital gown and wearing a protective mask from her chin to her worried eyes.

Saving the Best for Last

At the beginning of the decade, in April 2000, a new drug, Zyvox, and a new drug combination, Synercid, have been marketed specifically to treat highly resistant infections. Rushed through FDA approval, these new antibiotic therapies are the chemical cavalry meant to augment the threatened "drug of last resort," vancomycin, the current pinnacle of the antibiotic treatment pyramid. Vancomycin is still powerful enough to kill most serious bacterial infections, and few resistant bacteria float through hospitals. Even so, physicians wield such swords sparingly, using them only when a patient suffers from a serious infection resistant to other drugs. This is not because they're poor drugs—quite the reverse. It's because good drugs, if used widely, exert too strong a selective pressure on bacterial populations. Reducing the use of the best drugs—reserving them for the direst cases—serves as a first evolutionary strategy to combat antibiotic resistance.

In part because of this evolutionary control, resistant staph infections are typically still partially susceptible to vancomycin and can often be treated with high doses of that drug plus other (fairly invasive) means. However, warnings have gone out that staph resistance could increase and that vancomycin may have only a limited effective life span because other bacteria have more advanced resistance. Already, enterococcus bacteria, which don't cause staph infections but tend to grow inside digestive tracks and other dark places, have developed a high level of vancomycin resistance. Two genes, *vanA* and *vanB*, produce this resistance, but so far these genes have not shown up in naturally occurring *S. aureus*.

In this honeymoon phase, each strain of bacteria has yet to mature its own flavor of resistance to vancomycin. But scientists at a London hospital have recently shown that they could move vancomycin-resistant genes from enterococcal bacteria to staph. The chilling message stands out starkly: it happened in the lab, and it could happen in a hospital.

The evolutionary process couldn't be clearer in these cases. When we chemically chase staph infections away from our hospitals with a bull pen of major league antibiotics, we see the vanquished bacteria evolving. Accidental evolution manifests itself most seriously here, in the shift in bacterial diseases to circumvent whole series of lifesaving wonder drugs. But how do these simple bacteria successfully sidestep our wonder drugs?

CHEMICAL CONNOISSEURS OF the living world, bacteria have been on a liquid diet for 3.5 billion years, because, unlike more complicated cells, they have no internal cellular skeleton. Instead, bacteria rely on a stiff cell wall to maintain cellular shape, a cell wall that also prevents passage of solid food particles to the interior.

The first antibiotics worked by preventing proper construction of the critically important bacterial cell wall. The chemical sub-units that make up the walls act like bricks of a castle, and antibiotics like penicillin disrupt correct placement of the bricks so that the walls fail. Bacteria grown with penicillin make cell walls peppered with minute holes. Losing internal fluid like a breached oil tanker, these oozing bacteria quickly die.

Resistance to penicillin and similar drugs occurs when bacteria rebuild their walls, which they do in several ways. In one method, the protein used to build the bacterial cell walls has been altered to bind penicillin poorly, or not at all. Cells that make their walls with this modified protein can tolerate the antibiotics with impunity or can be killed only by massive doses.

A second, more common type of resistance depends on the simple tactic of breaking down the penicillin into harmless by-products. The protein these bacteria use to destroy penicillin has been co-opted by evolution into a new role, and by understanding how bacteria fuel themselves, we can see more clearly where antibiotic ability originates. Fed by chemicals in solution, and the occasional ray of light in some species, normal bacteria use their necklace of genes to make proteins that break down many different high-energy molecules for cellular fuel. A sweet example, the sugar glucose is burned by bacteria (and virtually all other cells) to power the tiny cellular engines of life. Another common example revolves around the beta-lactams. Proteins called beta-lactamases digest these high-energy carbon compounds in many bacteria, breaking them into smaller chemical kindling for the cell's metabolic fire. Penicillin resembles the beta-lactams chemically, having a peculiar circle of linked atoms known as the beta-lactam ring. When a bacterium encounters penicillin, the beta-lactamases inside the poisoned bacterium have the ability to destroy penicillin like they

TABLE 4.1. DATES OF DRUG DISCOVERY AND RESISTANCE

Drug	Discovery/ Introduction	Resistance
Penicillin	1928/1943	1946
Sulfonamides	1930s	1940s
Streptomycin	1943/1945	1959
Cephalosporins	1945/1964	late 1960s
Chloramphenicol	1947	1959
Tetracycline	1948	1953
Erythromycin	1952	1988
Vancomycin	1956	1988/1993
Methicillin	1960	1961
Ampicillin	1961	1973
Cefotaxime/ceftazidime	1981/1985	1983/1984/1988

would burn their normal beta-lactam fuel. If the lactamases act quickly enough, the bacterium might survive a mild dose of penicillin, being able to continue to make cell walls because the penicillin has been destroyed.

Other classes of antibiotics use different mechanisms to slow or stop bacterial growth, but in every case a bacterial response has evolved that thwarts antibiotic action. The common antibiotic tetracycline attaches to the ribosome, a part of the bacterial protein-making machinery, preventing protein synthesis by interfering with ribosome operation. Resistance has evolved by changes in the ribosomal genes so that the ribosomes of resistant cells no longer bind the drug. A second tetracycline resistance strategy has been the evolution of proteins in the bacterial cell membrane that can grab tetracycline from the cell's interior and dump it on the outside, where it can cause little harm. Like a bouncer in a late-night bar, these proteins quickly evict dangerous occupants before too much furniture is broken.

Virtually every antibiotic mechanism faces an evolved cellular defense that can operate to keep cells growing in the presence of the drug. For most drugs, resistance has evolved quickly after the discovery or widespread use of the antibiotic (table 4.1). Where do these mechanisms come from and how do they evolve? How does selection work in this case, and how could resistance to antibiotics have appeared?

The Engine's Roar

We live in a sea of bacteria. Some forge vitamin K in the bowels of our guts. Many more live ignored on skin and hair. Occasionally, bacteria enter the members-only club of our bloodstream through cuts and scrapes, or through nose or mouth. Infections occur when the population of a particular type of bacterium becomes too successful, breaching the body's immunological defenses. At that point, treatment with an antibiotic can slow this population's growth and help the immune system rid the body of the invaders. Antibiotics reduce bacterial growth rates, but what happens if bacteria in a population vary in their susceptibility to the antibiotic? In such cases, simple evolutionary principles predict that bacteria able to divide in the presence of antibiotics will make up an increasingly large percentage of the population. Bacteria that cannot divide will make up a smaller fraction as the more resistant cells outstrip them in growth. Eventually, most of the bacterial cells resist the antibiotic—this is simple natural selection in action.

How quickly does this natural selection lead to evolution? As we have seen, this depends on several factors. The rate of evolution depends on the amount of variation in a population in the trait under selection. For example, if beta-lactamases did not destroy penicillin and if there was no variation among bacteria in this trait, then beta-lactamase evolution could not occur. By contrast, prodigious variation in the way beta-lactamase acted on penicillin would allow evolution to proceed at a fast clip.

The rate of evolution also depends strongly on the strength of natural selection. Slight selection, such as that caused by a very mild dose of antibiotics, will cause very little change in the bacterial population. If, for example, a low dose kills only the most sensitive 5 percent of the cells, 95 percent of the population remains unaffected. By contrast, if a massive dose kills 95 percent of the cells, only those with the strongest resistance will survive, producing a marked change from the pre-antibiotic population. If 100 percent of the cells die during treatment, then selection halts completely because there is no population left to evolve.

Inheritance of resistance ability—in other words, the passing along of resistant traits to daughter bacteria—constitutes the final prerequisite of the evolutionary engine. Bacteria excel at this: reproduction usually entails a cell dividing into two genetically identical copies, with the exception of rare but important sexual events that resemble fax transmissions more than amorous encounters.

Where we find heritable variation in survival and subsequent reproductive success, we expect evolution, so the evolution of antibiotic resistance in disease bacteria should not surprise us much. But for bacteria at the dawn of the antibiotic age, where did the original variation for survival in the presence of penicillin come from?

A Bacterial Museum

British microbiologist E. D. G. Murray collected bacteria in the early twentieth century like others collected butterflies, gathering them up, mostly from human infections, and preserving them for future study and amazement. But Murray's approach was more like the modern day butterfly houses, which display winged rainbows in their live, fluttering glory rather than pinned in a glass and wooden box, insulted by mothballs.

The Murray Collection, started in 1917, consists of a microbial zoo of bacteria. The samples are archived in sealed glass ampoules in the British National Collection of Type Cultures. Few of the Murray bacteria have ever experienced industrial-strength antibiotics because they were collected from natural habitats, some before Fleming discovered penicillin. Nevertheless, among the hundreds of colonies stored in the collection, a few contain the genes necessary for low-level antibiotic resistance, especially to ampicillin and tetracycline. But why? Were these just the particularly well-prepared bacteria, with bomb cellars already dug before the war with antibiotics started? Because nature seldom predicts the future with any accuracy, the antibiotic-resistant cells of the Murray Collection must have occurred in other ways, probably for reasons immediately beneficial to the bacteria themselves.

First, some microbes make natural antibiotics for their own use in poisoning their immediate neighbors, and need an inborn way to protect themselves from their own weapons, producing proteins immune to the effects of their own antibiotics. The fungus *Streptomyces rimosus*, for example, produces tetracycline and has native resistance to it. The fungal genes responsible for resistance have been transferred to bacteria, where they give resistant ability to the bacteria lucky enough to have incorporated them. A second source of resistance ability may be the bacteria themselves, persistent targets of antibiotics produced by fungi like *Streptomyces*. Bacteria that would have been poisoned in the microbial game of chemical warfare may have naturally evolved mechanisms to detoxify the antibiotics created by their fungal enemies.

Whatever the reason, partial, naturally occurring resistance to antibiotics is the result. Moreover, the same mechanisms used to generate natural protection can act at least partially against some artificial antibiotics produced by humans, thus generating the evolutionary fuel for the rise of the first antibiotic-resistant diseases. Even if this variation was extremely rare to begin with—only 11 out of over 400 cultures in the Murray Collection showed antibiotic resistance—uncommon variants would eventually dominate once antibiotics became heavily and widely used. If in this new population of partially resistant cells there were any mutations for faster growth or longer survival in the presence of antibiotics, then the resistance genes themselves would evolve to be better and better.

This view puts our chemical war against the bacteria in a more dynamic perspective—it isn't a recent war, and it didn't start in 1943. It's a war perpetrated a long time ago by microbial combatants, a chemical battle of attack and defend in which fungal feints and bacterial countermeasures play out on a battlefield invisible to the human eye. Because the tiny combatants have already discovered the major chemical agents useful in subduing competitors, when we steal their ideas and make our own copycat antibiotics, plans to circumvent them probably already exist somewhere in the microbial world.

But another reason we face resistance of so many different bacteria has also been discovered; it explains why the spread of new antibiotic-resistance genes has been so rapid. In addition to chemical warfare, naturally occurring bacteria developed an ability we normally think of as human: long ago, in the microbial battle for chemical supremacy, bacteria invented espionage.

The Market for Secondhand Genes

James Bond, despite his ladies'-man reputation, never had the everyday bacterial skills of stealing detailed plans of chemical weapons during sex. The grubbiest bacterium, although without an ejector-seat sports car or the panache to wear a tuxedo on the beach, nevertheless is extremely skilled at obtaining novel weapon plans from another cell. Its moment of betrayal? That most intimate of bacterial encounters—half celebration of bacterial life, half desperate attempt to stay alive; the bacterial version of sexual reproduction—conjugation.

Conjugation exchanges bacterial genes through a temporary conduit that attaches two bacterial cells together at the waist—or where you might imagine a waist to be. Most of the time, bacteria reproduce by splitting into two identical cells with the same genetic composition, copying exactly the same single, naked chromosome as well as the supernumerary rings of DNA called plasmids. Plasmids, genetic pockets that hold small clusters of genes and other useful information that isn't usually included in the main genome, tend to house genes that confer survival ability in odd environments; they represent valuable commodities in the bacterial gene market.

When conjugation occurs, plasmids as well as copies of the whole genome exchange between interacting cells. The DNA strands pass through a tunnel between cells, carrying with them all the genetic information possessed by their former owner. The plasmids slip through too, thus allowing an exchange of antibiotic-resistance genes between different bacteria.

Although bacteria tend to conjugate with other cells of their own type, they are undiscerning partners and occasionally transfer genetic material with different bacterial species. In an alarming experiment, researchers at St. Thomas' Hospital in London were able to show how resistance to the antibiotic vancomycin could transfer to *Staphylococcus aureus* from other species—both in the test tube and in mice infected by multiple bacteria species. Although these experiments were highly artificial, they mimic other reported transfers of genes between bacterial species, especially when those genes confer strong selective advantage.

Conjugation is only one method of obtaining foreign genes. Bacteria like to try new things, and when swimming through their nutrient broth, they occasionally encounter bits of DNA debris from deceased bacteria. These random bits of genetic information can sometimes jump into the cell and if they survive their new intracellular environment, they can function to make proteins. Harmful genes imported in this fashion would kill the cell. But bits of DNA that confer antibiotic resistance could pay off during an antibiotic dose, so these bits of DNA are welcomed into the bacterial chromosome.

Because antibiotic-resistance genes can jump around, the evolution of this physiological attribute does not need to occur from scratch in every new infectious bacterium. Instead, the evolution of such an ability in one bacterial type can lead to rapid acquisition of the ability among many other bacteria, generating a serious problem in the use of strong antibiotics. Normally there is a lag time between the initial use of a new drug and the evolution of the first case of resistance. But once resistance has evolved somewhere, it can spread among species like wildfire. If each new bacterial target species had the same lag time, then we could switch a drug from treating one type of bacterial disease to another—from leprosy to staph to strep throat—and each time we would have a miracle drug for a short time. But the transfer of antibiotic-resistance genes from cell to cell means that once one bacterial species has developed resistance, resistance in the next species appears quickly. The lag time drops to near zero, and resistance becomes widespread—or universal.

Stern Nurses, Slim Hopes

So far, I have tried to describe the rise of antibiotic resistance from the perspective of the serious health risk it represents and from the perspective of the way our medical system has affected the evolutionary process. In one sense, heavy use of antibiotics demands evolution of antibiotic resistance as long as the evolutionary engine runs. But observing and understanding the evolutionary engine shows ways that medical treatment of infectious diseases can try to limit the engine's speed.

Reduction of antibiotic use is one method to accomplish this. Less use translates directly into less opportunity for selection, which (all else being equal) translates into slower evolution. Avoiding sublethal doses of antibiotics—by stringent attention to dosage recommendations—constitutes another evolutionarily derived treatment strategy, which has been effective in slowing the evolution of resistance. We learned this lesson when the world woke to a resurging "cured" disease, one that had been considered vanquished and relegated to medical school lectures about historic illnesses. By the 1990s, tuberculosis was back.

It had killed millions, from poets to princes, and until the antibiotic era, tuberculosis (TB) had no known cure. It killed by slow asphyxiation, or by so compromising the immune system that other infections became fatal. In industrialized urban centers throughout the world, TB raged throughout the nineteenth and early twentieth centuries. There were 135,000 cases of TB reported in the United States in 1945, many from poor areas in cities like New York and Los Angeles. But by 1985, TB had succumbed to the antibiotic bullet, and the number of new TB cases had dropped to 22,000 nationwide. A complex but effective treatment using two to four drugs cured it, and the attention of the medical SWAT teams turned to other pressing problems.

But in the 1990s, TB returned with a vengeance, with resistant strains becoming so resurgent that in 1993 the World Health Organization (WHO) declared a global TB emergency. In a 1998 report, WHO officials documented an alarming rate of drug-resistant TB: 10 percent of new cases resisted at least one drug, and it ballooned to 23 percent among people who had previously been treated. Multidrug resistance occurred in an additional 13 percent. Dixie Snyder and Kenneth Castro, writing in a *New England Journal of Medicine* editorial heralding the 1998 WHO results, point out that even in the 1950s, single-drug therapies for tuberculosis fell short because "the rate of spontaneous mutation resulting in resistance to streptomycin (and other drugs) is high enough so that a single drug cannot eradicate all *M. tuberculosis* organisms in persons with the disease." Thus, rapid mutation goes hand in hand with rapid evolution of drug resistance within a single person's infection.

BUT THERE IS another cause of widespread TB evolution: the uncritical way drug treatments are sometimes employed. Complete treatment kills *all* the bacteria in an infection, even those with slight resistance, causing the evolutionary engine to sputter to a halt, because the target population lacks variation for the ability to survive treatment overkill. But partial treatment opens the doors to evolution, since it allows some of the cells—those with a slight resistance—to survive.

TB progresses slowly; its cells divide at a leisurely rate of only once in twenty-four hours. As a result, the disease requires months of treatment for complete elimination. But why continue months of expensive medication when you feel much better after the first few weeks? Very often people will stop taking the drugs, saving money or giving them to family members, having dealt their infections a serious but not fatal blow. Worse, these people unknowingly have perpetrated a strong selection experiment among their infectious tenants, an experiment that could hardly be better designed to select for antibiotic resistance. This is because lesser doses actually select among variants in bacterial fitness, bringing the evolutionary engine roaring to life. Those few cells that have the best growth rates or survival skills during the initial treatment continue to grow and prosper as the treatment subsides. Therefore, TB evolves rapidly in an environment of partial treatment.

Although the problems associated with partial treatment apply to nearly all disease remedies, tuberculosis is an extreme case. In a Texas study in the 1980s, the rate at which TB reinfects a patient—the relapse rate—was about 20 percent. Many of the relapsed patients probably did not complete their course of medications. But the study also followed the fate of patients who were required to take all their pills, stared down by stern nurses with no-nonsense shoes. This treatment method, called direct observation therapy, which was invented in Tanzania in 1977, dramatically reduces the rise of antibiotic-resistant TB. For this group in the Texas study, the relapse rate, and the rate of development of antibiotic-resistant strains, dropped to about 5 percent. After implementing direct observation therapy in New York City, new cases of TB that were resistant to one or more drugs declined 80 percent from 1992 to 1996.

REDUCTION IN FUNDS for tuberculosis treatment under the delusion that this disease has been conquered (or sometimes due to the collapse of a national economy) can lead to unchecked transmission of resistant infections, often through the agency of partial treatment. And the evolution of any disease makes its treatment slower, more difficult, and usually more expensive. Treatment of antibiotic-resistant strains of TB can run $10,000 a case, a figure that breaks the bank in many developing countries. By some estimates, contracting an antibiotic-resistant infection during a hospital stay can increase hospitalization costs by $9,000 for a methicillin-resistant strain. Costs can skyrocket by over $100,000 for a vancomycin-resistant infection. Overall, antibiotic resistance adds an estimated $30 billion to the annual U.S. health bill.

Fully one third of humanity now carry TB, and over 3 million die from it each year. The huge number of new antibiotic-resistant cases has left some with an infectious TB so far incurable, and left millions of others in need of advanced, impossibly expensive medical treatment. Thus any existing political and economic problems become exaggerated by the evolution of antibiotic resistance, and sometimes a disease evolves so far that, although in theory curable medically, it can't be done because of economics. Unless society can conjure up a medical fairy godmother to produce cheap and widely available antibiotics, the evolutionary cycle has also ratcheted up the social and economic costs of this disease.

When the Enemy Evolves

The evolutionary wheels turn and turn in individuals taking simple treatment for diseases like TB. Terrible laboratories for the evolutionary experiment, each turn of the resistance cycle makes the disease more and more difficult to treat. TB represents the first threat here, but we can finger the second culprit as the evolutionary process itself, which is vastly more difficult to influence. In fact, TB presents a model case of a reinvigorated medical community beginning to integrate the fact of evolution into practical treatment strategies.

New guidelines include strategies that work to limit evolution among all carriers of the disease. For example, a list of treatment suggestions from the New York City Department of Health includes many that take a solid evolutionary approach to combating TB. It recommends:

1. "Always test a patient for drug susceptibility." This allows individual treatment with effective drugs, avoiding broad-spectrum drugs that can lead to the evolution of broad resistance.
2. "Always treat with a regimen of at least two—and preferably three to five—drugs." This dramatically reduces the variation in sensitivity among the bacterial population—they'll all die—and lowers the chance that a new, single mutation will arise to revitalize the bacteria. Without variation in ability to grow in the drug cocktail, bacterial evolution virtually halts.
3. "Never add a single new antibiotic to a failing drug regimen, always add two or three." Again, a failing regimen points to the presence of resistant cells in the infection. Adding just one drug to the list would, in effect, initiate new treatment with just one drug—and boost the chance of evolution.

4. "Treat patients who have multidrug-resistant strains for at least eighteen months." This will prevent lingering, partially resistant cells from thriving at the cessation of a treatment period too short to actually kill them all.
5. "Always treat multidrug-resistant patients with direct observation therapy." This observation ensures the patients have taken the full course of drugs, thereby reducing the number of cases in which resistance can arise because of partial treatment.

Of course, attacking any disease is more than a purely evolutionary problem, and diseases like TB can present as many political problems as medical ones. In 1993, for example, health department officials in New York City added detention to their list of tools for combating TB. People who had had TB multiple times, who refused treatment and left hospitals against medical advice, were detained against their will in locked hospital wards until the disease had been fully treated. Of the 139 people who were "jailed" for failing to treat their disease from 1993 to 1995, three quarters were cured (most of the rest died of unrelated diseases like AIDS). Such draconian measures, as well as the aggressive treatment of TB patients with direct observation therapy, have been credited with short-circuiting the threat of TB epidemics in Seattle and New York City. However, as Barron Lerner's *Contagion and Confinement* documents, taking away a citizen's civil rights for his or her own good is a dangerous option for societies to exercise. Lerner describes the detention of 2000 people at Seattle's Firland Sanatorium between 1950 and 1970—most skid-row alcoholics—for involuntary treatment of TB. In New York, officials point out that only 2 percent of the TB patients there were confined and assert that their program balanced societal and individual needs.

Failure to treat yourself correctly for a contagious disease may result in evolution of more dangerous pathogen, but does this mean that individuals must give up the right to decide their own treatment plans, or that the protection of society by reducing disease evolution is the paramount concern? If it were, you might expect detectives rooting around in your medicine cabinet to make sure you took all twenty penicillin pills for your last infection, and to haul you off to jail if you didn't. And if failure to reduce evolutionary rates were a crime, then there would be bigger criminals than skid-row alcoholics.

The Bacteria Get Bossy

As we can see, sensible modern TB treatment strategies are the fruit of the evolutionary tree. But that fruit is not as ripe as it might be, because evolution-reduction strategies have not been evenly applied: drug use in humans may be under control, sometimes at the cost of confinement, but use in animals is not.

Never having seen a caribou in the doctor's office for a chest cold, you might reasonably think that humans must constitute the major antibiotic market. However, livestock farmers, in growing a huge number of animals for human food, purchase 25 to 50 percent of antibiotics used. Moreover, the two simple recipes for reduced evolution—judicious drug use and full, curative treatments—fail when livestock farmers generally ignore them in their efforts to increase growth and protect against disease in their stocks.

In industrialized countries, farmers have long since learned that antibiotics increase weight gain and decrease infections. In many cases, antibiotics constitute a standard part of animal feeds and are used prophylactically to prevent disease outbreaks. Naturally, these prophylactic uses call for smaller doses than one might use to treat an active infection. At one fell swoop then, both of the resistance-resisting treatment strategies outlined above collapse. One might worry only a little about this—such treatments must work well for the farmers or they wouldn't use them. And surely few livestock live long enough to develop full antibiotic resistance in their internal bacterial cargo. True enough. But unfortunately, many resistant bacterial strains that develop in livestock prove impossible to keep on the farm. Instead, they break out into the countryside looking for their next big mammalian host—in all likelihood, that host will be us.

One of the deadliest disease bacteria—the strain *Escherichia coli 1057:H7*, or *E. coli* in newspaper-speak—probably originated in livestock. On that circular bit of DNA known as a plasmid, added as a hitchhiker to a normal bacterial chromosome, *E. coli* contains genes for a deadly toxin that kills host cells on contact. First noticed in 1982, in contaminated meat, this killer has been found in undercooked meat, in drinking water supplies in New York and Missouri, in apple cider (from contaminated manure used as apple fertilizer), and it can easily lodge as a fatal infection in otherwise healthy people. It has also recently evolved multidrug resistance.

Even just a few new cases of *E. coli* represent a medical emergency serious enough to demand closure of suspect meat-packing plants. But although such facilities may foster these infections, antibiotics in the feedlots on massive farms and in solitary barns generate the evolutionary gale that pushes this disease ashore.

Nor is the global livestock connection restricted to cattle. All over the world, antibiotic use goes with livestock rearing—in fact, in some countries chicken or beef persistently show antibiotic loads higher than FDA tolerance limits. So many animals have been chronically treated with antibiotics in country after country that meat even from healthy animals shows underlying bacterial contamination by resistant strains. Inspectors have commonly discovered *Salmonella*, for example, among chicken carcasses sampled from processing plants. When isolated and tested for antibiotic sensitivity, these *Salmonella* show strong resistance to amoxicillin, tetracycline, streptomycin, and many other antibiotics, and can cause drug-resistant diseases in humans. They can also transfer their resistance genes to nonresistant *E. coli* that thrive naturally in the human digestive system. The *International Journal of Food Microbiology* recently reported that many chickens delivered as food to hospitals in Greece had *Salmonella* contamination, most of which were resistant to one or more antibiotics.

Lady in Red

Antibiotics generate a pounding evolutionary pressure. Bacteria evolve in response, to attain thick-skinned resistance against chemical attack. We shift culturally and industrially, altering the chemical mix of antibiotic treatment. The disease dance continues, turning to the evolutionary tune, and both players must step smartly. This cycle—played out on the evolutionary stage many times before—is called co-evolution, the simultaneous evolution of species to become further adapted to each other.

Such evolutionary escalation and counterescalation runs rampant in the natural world wherever strong and consistent interactions among species exist. Predators of the African savanna, for instance, continually shape the legs of their prey. The fleetness of a cheetah justifies the speed of its antelope quarry and vice versa.

As first noted by Lee van Valen at the University of Chicago, this evolutionary cycle generates a kind of co-evolutionary arms race, dubbed the Red Queen hypothesis, after the Red Queen in Lewis Carroll's *Through the Looking Glass* who runs all day just to stay in the same place. The Red Queen hypothesis seems to govern many natural interactions between predators and prey, or between hosts and disease. Could these natural examples shed any light on our current arms race with infectious, antibiotic-resistant bacteria? Better success on our part in poisoning bacteria leads to selection on them to overcome our drugs, which results in stronger pressure on us to develop a new strategy. That the bacteria evolve via normal genetic mechanisms yet our strategy rests upon cultural and industrial shifts does not change the basic ebb and flow of the interaction. Our greatest need is constant updating of our response to infectious bacteria and constant monitoring of the bacterial response. We should not consider evolution of the bacteria as an unlucky disaster, but instead recognize it as the expected evolutionary outcome, one we can predict as an understandable evolutionary consequence of disease treatment.

Darwin Meets Machiavelli

With such high stakes we need to avoid glib evolutionary pronouncements about the best public health strategies. But evolution might offer a few new avenues to explore, and provide ways to learn from the natural world's look-out-now ecological diversity. What strategies suggested by an evolutionary perspective might successfully interrupt the steady generation of antibiotic resistance?

First and foremost, an evolutionary perspective paints evolution of drug resistance as the baseline, *expected* outcome of any long exposure to a drug-based treatment regime. Instead of ignoring the evolutionary engine, we must seek to understand and perhaps undermine it.

For example, reducing variance in fitness offers a powerful way to slow evolutionary change. Strategies already incorporated into the antibiotic treatment protocols call for multidrug therapy or drug overkill to eliminate all cells in an infection. Under this scheme, if all cells die, variation in fitness drops to near zero, and evolution ends with it.

Other possibilities arise once we admit a world controlled by evolution. For instance, can the evolutionary engine be reversed? Can we use the evolutionary process to our advantage? What happens to antibiotic resistance when drug treatment stops? If maintaining the ability to destroy antibiotics demands a high cost, then, once the antibiotic treatment stops, evolution will cause antibiotic resistance to disappear as quickly as it appeared.

A well-known example is resistance to tetracycline, which requires evolution of an altered ribosome, one that resists poisoning by the antibiotic. But these very ribosomes make proteins more slowly than do nonresistant bacteria—a trade-off that selects for the original, nonresistant cells when tetracycline levels drop. In cases like this, when superiority of resistant cells depends on the drug environment, resistance swings like an adaptive pendulum.

Other bacterial adaptations that confer resistance also embody strong costs. Plasmids that carry antibiotic-resistance genes inside cells may be small, but they require time and energy to duplicate during cell division. This cost, like extra luggage on a crowded plane, slows down the bacteria. Bacteria without extra plasmids are out of the airport and having a mai tai on the beach while the plasmid-bearing ones are still waiting for their last genes to be delivered by the baggage carousel of cell duplication. This means that cells without plasmids, which grow well in the absence of antibiotics, may have higher growth rates than those with them.

In some cases, antibiotic-resistant cells do seem to pay a sort of metabolic cost for their resistant abilities. Careful work on experimental bacterial mixtures has documented a 10 to 20 percent disadvantage compared with normal cells lacking the extra resistance options. Unfortunately, we do not always observe such costs, and clearly some bacteria can adapt by reducing them. Work by Bruce Levin and colleagues shows strong selection for bacteria to reduce the costs of antibiotic resistance. In some cell lines, bacteria so successfully maintain low-cost resistance ability that they can persist for 10,000 generations in antibiotic-free cultures without losing to antibiotic-sensitive competitors. So far, this bodes poorly for engine reversals. But low-cost resistance mutations come along only after the original ones have spread. Perhaps we will discover a general rule that the first cells inventing resistance to a new drug will show this fitness cost and be sensitive to reversing the evolutionary engine.

Another suggestion comes easily from the natural world of coevolution: other bacteria may combat our diseases for us, possessing strategies to limit growth of the cells we deem harmful. Competition among bacteria, studied so extensively in the lab, is now gaining appreciation in nature. Can normal levels of benign bacteria bar colonization by harmful ones? Perhaps antibiotics, which wipe out bacterial friend and foe alike, reduce competition among bacteria and make it easier for harmful strains to reinvade. In 1899 Ilya Metchnikoff recommended that dysentery be treated with active cultures of lactobacillus. Called replacement therapy, this approach sought to replace a harmful bacterial infection with a benign one. Some modern studies have shown competitive effects of established gut bacteria on those trying to invade, and reports in 1996 suggested we could protect chickens from *Salmonella* by an inoculation of several dozen other bacterial species. Nevertheless, very little work has yet been done on the interactions among friendly and harmful bacteria. Whether this avenue offers any hope in the antibiotic arms race has yet to be seen.

WE HAVE COME a long way since the industrial debut of penicillin in 1943. Resistance now affects every drug that currently exists, and it will continue to operate on all drugs in the future. The lesson that percolates here—that evolution has a practical face—is written in the rapid shift of features, like antibiotic resistance, that matter to the daily lives of us humans. This means that evolution is no mere curiosity of the natural world, but a potent process that we must understand. Not only is evolution the fundamental principle around which we can make sense of the complex biological world, it has teeth too.

Chapter Five

The Evolution of HIV

IV is a personal disease—not just in its transmission, or the hurtful stigma of the ill informed, or even the mortal intensity of its rampant destruction. It's personal because no two infections are ever the same. Each generates a private nemesis, a unique battle at the gates of an individual's immune system. Though branded with the same name, consequences, and prognosis, every HIV infection evolves as a unique, irreproducible event; it's a trajectory rather than a disease, more a serial drama than a caged statistic.

An HIV infection evolves so fast that within two months it has hidden itself from the alerted immune system. If identical infections start on the same day in two different people, they immediately begin to diverge as they individually cloak themselves in grim disguises, using different evolutionary paths to short-circuit each person's stunning but short-lived ability to destroy the first viral wave. The ability of the virus to disguise itself, powered by the evolutionary fuel of genetic mutation (a fuel that burns like a rocket mixture compared to coal), makes HIV a contender for the title of fastest-evolving entity we know.

The mutation rate of HIV underlies its success like a good smile underlies a political career, fueling such rapid evolution that it happens *within every person who contracts the disease.* Within two years of infection, HIV can evolve from a genetically homogeneous virus to a mixture of viruses more different from one another than humans are from chimpanzees. These newly evolved virus types travel under the same name, HIV-1, and they have the same basic modus operandi. But they are an invading army of tanks, missiles, infantry, and amphibious troop carriers, all evolved from the single foot soldier that originally entered that body.

In 1999, HIV infected more than 40 million people. Among viruses, only the influenza epidemic after World War I has killed more people. An evolutionary battle, our war with HIV includes three main theaters of action: a high mutation rate, selection on viruses to evade the immune system, and selection to resist antiviral drugs. By nature, HIV infects the cells that coordinate our immune defense, thereby damaging our ability to fight off the infection. This attack on the border patrol of our bloodstream causes the collapse of the immune system itself, ushering in the hegemony of HIV. Clearly, to fight HIV we must ultimately conquer its ability to evolve.

How HIV Works

Old clocks possess a confidence shown in every chime. My living room houses one of these antique timers, left with us by my sister-in-law, and from its central vantage on the mantelpiece, it sits, silent through most of the days, waiting for her return. Every few months, as if walking a mechanical dog, I wind it up and let it run through its chimes, stretching its hands in a run around the dial, calling out the hours of a Massachusetts evening.

During those hours, the craftsmanship of the clock's original maker awakens. Written in steel gears and ribbon springs is a mechanical plan that counts seconds into minutes and hours into music. I know when I hear the chimes that they are connected deep inside to the workings of the clock—and that the tiny events of gear teeth and oscillating counterweight transform with mechanical precision into the quiet punctuations of the hours. The clock runs because its machinery faithfully traces the mechanical vision of the clockmaker, from the hands all the way inside to the smallest, coordinated spring.

Other, biological, machines are also driven by central mechanisms, though they are constructed by a wet and pliable machinery of living components. Each individual organism is like a clock with a clockwork inside. Among the simplest of these organisms are the viruses, and here the connections between life's machinery and its function show clearly. A virus's central machinery is its genetic information, or genome, the chemical plans that (like the gears of our clock) set a virus in motion, defining its potential to activate and grow. A virus needs only its metabolic springs wound to begin ticking.

HIV has a genome, as do all living things, and in that genome lies written the potential of the virus to mark the hours of a person's demise. The genomes of most organisms, even the simple bacteria, are constructed with DNA (deoxyribonucleic acid), but the genome of HIV consists of building blocks made of RNA (ribonucleic acid). This feature qualifies HIV as a retrovirus, and by enormously affecting HIV evolution, sets it on its killer course.

RNA differs from its famous cousin, DNA, in having an extra

hydoxyl group, a mated pair of oxygen and hydrogen, attached to a ribose (a particular sugar) that forms the chemical ribbon of the genome. This minor chemical change has profound consequences because it causes RNA to be much less stable chemically than DNA. We used to think of genes as unchanging, thanks partly to Gregor Mendel's experiments, but we now know that long nucleic acid chains like DNA and RNA are heir to many environmental insults. Even simple water attacks it, and at normal mammalian temperatures and cellular acidity, water chemically damages small parts of every gene every day. Like the lightbulbs in a downtown marquee, genetic components slowly flicker out, needing constant replacement to sustain their message.

DNA, the acid-free paper of the genetic world, stably nurtures its information in a way that limits environmental decay. This stability augurs well for its use as blueprint material, and DNA is used by most organisms to store genetic data. By contrast, RNA falls apart spontaneously at a much higher rate. Perhaps as a result, in cellular life-forms like humans, RNA occupies only temporary jobs like carrying information from the nucleus to the cytoplasm or (with stern protein chaperones) translating this information into proteins. RNA is so unstable that only a small number of viruses use it as their main, long-term repository of genetic information. However, these so-called retroviruses tend to be trouble when they develop a taste for humans.

Viruses that have retained their RNA genome, like HIV and influenza, are burdened with a high rate of chemical degradation, leading to a high mutation rate. Mutation in HIV outpaces even the other retroviruses, because HIV is saddled with a genetic copying protein that makes a large number of mistakes. As this protein, an enzyme known as reverse transcriptase, plays its key role in the life cycle of HIV, the growing retrovirus accumulates genetic changes in nearly every gene, providing raw material for evolution in most of its parts.

Family Tree

HIV has been around long enough, with its high mutation rate, to accumulate a zoo's worth of genetic differentiation. But, AIDS's sudden onslaught suggests that HIV is a virus new to human history. When did HIV arise and where did it come from? Immunodeficiency viruses are common in other mammals, both wild and domestic, so attacking mammalian immune systems is an old viral lifestyle. However, in other mammals, such viruses differ strongly from HIV. The one exception is the viruses that our cousins, the apes and monkeys, contract.

Different viruses tend to reside in different host species, and are as loyal to them as frequent flyers are loyal to an airline. As a result, evolutionary relationships among virus genes tend to mimic those of the hosts themselves, so that the immunodeficiency viruses of monkeys and other primates resemble one another more than they resemble, say, cat viruses. In particular, gene sequences in HIV most closely resemble sequences in the simian immunodeficiency virus (SIV) and are very distant to feline viruses. Surprisingly, though, HIV falls onto the virus family tree in two places. One form, called HIV-1, causes the current HIV epidemic in humans, and is most closely related to viruses from African green monkeys and chimpanzees. Another form, HIV-2, falls in a different place on the tree, is largely restricted to Africa, and produces a milder form of infection. It resembles more closely the SIVs from other monkeys, like sooty mangabeys. Differences in these viruses led them on very different tracks: HIV-1 causes the collapse of the host immune system, the condition we call AIDS, acquired immunodeficiency syndrome; HIV-2 does not cause AIDS; HIV-2 has a more faithful reverse transcriptase enzyme and a lower mutation rate, and it resides more timidly as a nonlethal parasite in human immune-system cells.

The similarity of immunodeficiency viruses to one another suggests one ancient source—perhaps a retrovirus that long ago mutated to prey upon the immune system of an extinct primate ancestor. Somehow the progeny of this viral inventor has moved from one species to the next, ever expanding its range of hosts. But despite broad similarity among viruses, each primate tends to harbor its own unique type—one that works best within that species' immune system. These important differences among modern viruses in their various host species tell us that viruses do not switch very often among hosts. If a lot of indiscreet virus jumping took place, the same virus would spread widely among different animal species, and humans would harbor SIV as well as HIV.

Recent genetic detective work has nailed down the place where HIV-1 jumped into humans. Among simian immunodeficiency viruses, it is perhaps no surprise that the one from chimpanzees is the closest to HIV, since we know chimps most closely resemble humans genetically. But chimpanzees are a widely distributed, complex species, with several subspecies that live in different parts of Africa. Culturally and genetically different, these subspecies have diverged recently. Only one of them, *Pan troglodytes troglodytes,* from west central Africa, was the source of HIV.

The evidence for this conclusion lies in the genome sequence of the SIV from *Pan troglodytes troglodytes.* When compared with the HIV genome sequence, and the sequences from other chimp subspecies, the *Pan troglodytes troglodytes* sequence is the closest match to the particular type of HIV, type N, that occurs endemically in humans in west central Africa. Genetic evidence also shows that three different types of HIV invaded humans about the same time from chimpanzees. Chimpanzees were hunted for food in west central Africa, and the accidental transfer from chimp blood during butchering could easily explain this movement.

When Did HIV Arise?

Transfer from chimps to humans was just the beginning. Once HIV-1 arrived, it began a riot of evolutionary diversification with a path we can trace in the genetic signatures of modern virus genes. Since the invasion of humans by HIV-1, a half dozen major HIV strains have evolved, accumulating a Pandora's box full of genetic differences. Fueled by rapid mutation, about 20 percent of the RNA genome has changed among HIV strains; this is a genetic diversification that exceeds that found between dolphins and whales, or between humans and baboons. How quickly this diversification occurred remains a matter of conjecture, but HIV genes can be traced back to an ancestral virus whose genetic composition has been reconstructed as a sort of "average" among all existing HIV strains. When this original type existed is still uncertain, but one very much like it has turned up in cold storage.

In 1959, 1213 blood plasma samples were taken in Africa. In 1997, these stored samples were rediscovered, and one from the Belgian Congo tested positive for HIV. Gene-sequencing studies subsequently showed that this individual harbored a version of the HIV-1 virus amazingly similar to the predicted ancestral sequence deduced from the modern HIV-1 strains. Although not a perfect match, these data suggest that the person who had this virus in 1959 contained a version very close to the HIV-1 ancestor. Thus, the shift of HIV-1 into humans must have taken place before 1959, but probably not too long before.

Since then the virus has been an evolutionary fire, developing into a blaze of dangerous strains that can confound drugs. New strains emerge frequently that can sometimes differ so greatly from common strains that diagnostic tests for HIV can miss them. Different strains dominate various parts of the world, strain B being the common form in Europe and North America. Central Africa harbors the widest mix of types, with types A and D common there, along with a rampaging type C that appears to be adapted to heterosexual transmission. Type A gained ascendance in west Africa and type C rules southern Africa and India. Thailand suffers from a major outbreak of type E, whereas in Brazil type F abounds.

The Eye of the Virus

Look at your credit card, its color, and the glittery picture on the upper right corner. Maybe that hologram has amused you in an idle moment. You recognize the name of the bank that issued the card—maybe since they are so good about correspondence. All those things constitute the morphology of the card—the sum of its physical features; some are window dressing, others are important. But the major information content of the card lies embossed on the raised numbers and buried in the magnetic stripe on the back—the card will not operate without these codes.

A virus resembles your credit card by having the same contrasting traits, an exterior set of physical features that herald the virus identity, with the real information content buried deep inside.

The HIV virus appears (figure 5.1) as a starburst globe of protein decorated on the outside with triple-decker studs of specialized proteins. The studs consist largely of two proteins called gp120 and gp41, which serve to latch the virus onto cells so the

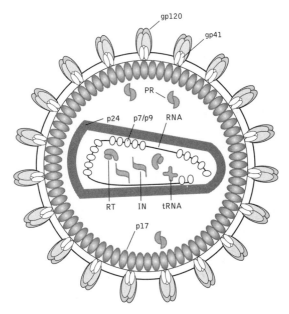

Figure 5.1: A drawing of the HIV virus showing the exterior studs made of proteins gp120 and gp41, the outer envelope proteins (p17) and the interior time capsule of the RNA genome.

virus can then invade those cells. Under the studs lies a protein envelope defining the virus's shape and providing a cinder-block-like foundation supporting the gp120 and gp41 latches.

Underneath this layered surface, at the center of the globe, lies a capsule containing the viral genome. Wrapped and secure, the RNA strands inside this vault lie quiescent and safe, full of deadly potential, waiting to be injected into a new, naïve host cell. Lined up along the RNA ribbon of the genome are the few genes absolutely necessary for the virus to succeed—the ones that coordinate the attack of HIV.

Like a terrorist squad, the genes of the virus lay waiting for delivery into a sensitive cell. They are programmed to infiltrate the cell's nucleus and commandeer crucial cellular functions. Once inside, the virus genes switch on and subvert the invaded cell's metabolism, slaving the cell to produce more virus. The infected cell leaves a legacy of more virus particles, creating the next generation of cellular subversion.

A Day in the Life of HIV

Jeffrey became a friend of my entire family, but he was first an AIDS patient of my wife, Mary. He would appear in his canary yellow, broken-down Rolls-Royce, accepted in lieu of cash in some scam years before, laughing loudly and ignoring the many dents and rust spots—on his car and on his life. The car defined Jeffrey's mix of style and poverty, making sure you never knew if he was riding on top of the world or just doing the valet parking. Either way, he thrived on friendship and stories, bursting with energy—when his disease was under control. Jeffrey's life, like everyone with a growing HIV infection, was lived on two levels: the exterior, where daily events could be completely normal, and the interior, where an unseen struggle raged. Our friendship spanned that struggle, but none of us except Mary knew a lot about it, because the evidence of the immune wars could at first be read only by subtle internal clues. But what happened to Jeffrey was what happens to all people with AIDS, in a disease trajectory driven by the battle plan of HIV.

HIV enters the body ready to invade a host cell. But HIV is pickier than a cat that isn't hungry, and there are only a limited number of potential targets. Circulating in the bloodstream or the lymph nodes, HIV can attach to only a few types of cells. Other viruses are more catholic in diet, infecting cells that line your nose or lungs, creating head and chest colds. HIV invades only white blood cells, and even rejects most of those, all but the very few that display on their surfaces a particular protein called CD4. These cells, called helper T cells, function as part of the human immune system. So instead of attacking the cells lining your nose, HIV attacks the cells coordinating your immune defenses.

The proteins on the outside of the virus attach to CD4 receptor proteins on the helper T cells (figure 5.2). After attachment, HIV drills a hole in the T cell membrane, providing an entry point for the viral RNA genome. The RNA unwinds from the capsule at the virus's center and slips into the host cell like a burglar through a skylight. Once inside, the RNA immediately begins to work. Taking over a team of cellular ribosomes, the viral genome tricks the cell into making a set of virus proteins from the virus's genes. Of the viral proteins made at this stage, the most dangerous are reverse transcriptase and integrase.

Reverse transcriptase copies the viral RNA strand, pilfering raw materials from the cell. But instead of making a duplicate RNA strand, reverse transcriptase makes a DNA copy, which mimics parts of the cell's actual genome but contains all of the viral genes. Because they appear similar to the host's DNA, the cell brings the viral genes into the nucleus and inserts them into the normal chromosomes. Here, the viral integrase protein renders a critical assist by integrating the viral genome innocently into the genome of the host cell.

Once inside, the viral genome can have two trajectories. Like Rip Van Winkle, the virus DNA can sleep for a long time before becoming active again. The more common scenario, however, is to become active and deadly. Most viruses are not content to enter and hide. They enter and kill.

Killing requires further cellular subterfuge—a complete annexa-

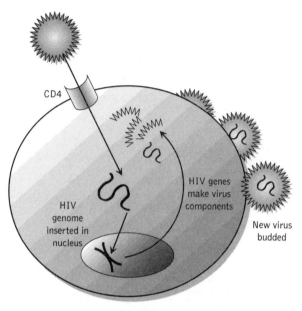

CD4

HIV genes
make virus
components

HIV
genome
inserted in
nucleus

New virus
budded

Figure 5.2: HIV enters a white blood cell through the CD4 receptor, and the RNA of HIV genome is incorporated into the host cell's chromosomes. Once there, the viral genes make the pieces of new viruses, which are then assembled in the cell and are budded from the cell surface. Once free, the new viruses can infect other cells.

tion of the cell's functions by an attack squad of mobilized viral proteins. The integrated viral DNA, still masquerading as a regular scout troop of cellular genes, uses the cell's normal machinery to construct the set of proteins needed to make more virus. The proteins include the gp120 and gp41 exterior studs, as well as the other structural components of the viral coat and the RNA capsule walls. A protease, a potent viral protein jigsaw, cleaves proteins into the precise components needed to make new viral coats. Finally, these coats are assembled on the cell surface, forming protein globes that will house the virus's RNA strand. Then the hidden viral DNA, nestled permanently in the cell's nucleus, engineers one last betrayal: it causes the cell to construct a horde of RNA copies of the virus genome. The RNA rats scurry into the safe havens of the waiting viral globes—one to a lifeboat—the hatches slam shut, and the globes launch into the bloodstream. A new generation of virus, born from the slavery of the overwhelmed T cell, spews into the body of the host, where the HIV begins a new search for a new cell and a new cycle of destruction.

HIV War I: The Battle of the Immune System

The invasion and destruction of the first unlucky CD4 cells inaugurates an HIV infection, beginning the HIV war against the human immune response. But the initial invasion does not go unnoticed by our immune system, a flexible and responsive defensive network of cellular smart bombs that is perfectly capable of seeking out and destroying HIV viruses—at first. The rapid reaction of our immune system to HIV infection sets the stage for the beginning battle and causes the first of the evolutionary pressures acting on HIV.

HIV may be a new virus, but viruses in general have been around for a long time and the human immune system is particularly well designed to vanquish them. A typical immune attack has more plot twists than a soap opera, but the overall interactions are fairly straightforward.

HIV virus particles are attacked by indiscriminately hungry white blood cells called macrophages. Although this assault is not enough to affect viral numbers seriously, it begins a cascade of events that usually leads to a powerful immune response against the invading virus.

The macrophages envelope the virus and signal their victory by displaying on their surfaces a small trophy of the encounter—a piece of the HIV virus itself. Helper T cells each recognize a single, specific protein trophy on their neighboring fellow white blood cells. Among the many types of helper T cells, those with a CD4 receptor specifically recognize macrophages that have swallowed an HIV virus.

Once they encounter each other, the CD4 helper T cell binds to the HIV-trophied macrophage, releasing a hormone called a chemokine, which in turn ignites two types of critical cellular responses. In the first, white blood cells called B cells release antibodies that attack the HIV viral particles. In the second, killer T cells are activated that destroy all cells displaying the same trophy as the one that activated the helper T cell. The antibodies bind to free viral particles before they have a chance of infecting cells, wrapping them in a coat that allows their easy disposal. By contrast, killer T cells rid the body of any cells that have already been infected, because infected cells display the trophy proteins as a sort of "I'm Infected, Stay Away" sign that targets them for destruction.

The Cavalry Stumbles

Buried in the core of this immune success are the seeds of immune-system failure. Imagine the situation right after HIV infection. The virus attacks its first cellular victims, CD4 helper T cells, destroying them in order to reproduce more virus. Macrophages encounter and swallow the circulating virus, sparking the response from B-cell antibodies and killer T cells. By turn, the killer T cells target and destroy cells infected by the HIV virus—the very CD4 cells that recognize HIV! So, CD4 cells are killed by HIV and killer T cells, a double whammy that quickly depletes their numbers. Victory over the virus, rather easy at first, occurs at a high cost: that of CD4 cell death, and future ability to recognize HIV.

If this were the end of the story, then HIV would be merely a bad disease. Once activated, the antibody and killer T cell defenses could scrub the system clean. The demise of the CD4 helper T cells that started the immune response would be tolerable, because the immune response would have already begun. If a smoke alarm burns, it has still done its job—as long as it called in the firefighters.

But HIV wins the viral prize as a master of disguises, a master that can change its outer face and still remain HIV at the core. It's as if the firefighters arrive at the scene to find that the fire has mutated into an unrecognizable threat, a smoldering, lurking menace that no longer triggers the smoke alarms. The firefighters would encounter a burnt scene and would extinguish any fire they found, but they'd go away without realizing a different threat remained behind.

This unrecognized threat stems from the mutated versions of HIV. These mutants are resistant to attack by antibodies and are ignored by the first wave of killer T cells. The mutant disguises are effective because the immune system recognizes only a few proteins on the exterior of the viral globe—particularly the gp120 and gp41 proteins—and if this exterior mutates, then the original HIV antibodies and killer cells can no longer attack them.

Mutation generates enormous variation in HIV proteins, including variants that can disguise the virus from the first wave of immune defenses. Mutants survive the onslaught of the first immune defense, then begin seeking new cells to invade. Although these viruses do start a new round of immune-system recognition, antibody formation, and killer T cell activation, these new viruses are king for a little while. At this point, the HIV infection might as well be starting afresh in a naïve host, infecting another set of CD4 helper T cells and setting off another round of frantic antibody and killer T cell production. This time the immune attack keys on a slightly different viral disguise, but there are fewer CD4 cells around to coordinate the response.

HIV evolves in this stepping-stone fashion, responding quickly and incrementally to the immune system, an evolutionary drama that continues as long as the immune system functions. This co-evolutionary cycle spirals downward as the immune system generates selection on viral mutants, which then cause a new immune response to the new virus. Like the Red Queen, both players must run all day to stay in the same place. However, the immune system cannot run all day. Every *immunological* victory over HIV develops a new set of killer T cells that destroys a new set of helper T cells. Every viral mutant quelled gives rise to a brood of new mutants that successfully invade helper T cells, and every cycle sees the immune system running more and more slowly, with fewer and fewer helper Ts coordinating the next immune attack.

Like a well-trained army, the immune system has the firepower to conquer HIV. But outnumbered by viral mutants, the system succeeds itself to death. The result—eventual collapse of the immune system—occurs when HIV has so seriously depleted helper Ts that the immune system can no longer recognize an HIV infection. This stage of HIV infection heralds a huge increase in viral loads in affected individuals. Unrecognized by the immune system and unimpeded, HIV then proliferates wildly. At this point, the immune system flails too weakly to fight off other types of infections—and AIDS sets in.

———

Being Hawaii

Evolution usually strides the widest corridors of the biosphere, precipitating global biological changes with sweeping import. Dinosaurs were once on evolution's playlist, dominating the land, sea, and sky. Today mammals and flowering plants are ascendent, and few terrestrial habitats lack them.

But evolution has a quiet side too, a side that produces boutique species in cul-de-sacs. I have stood in the bowl of Koko Crater in Hawaii and dug in the surprise mud of a torrential rain to find a tadpole shrimp that lives nowhere else on Earth. This unique species, molded by the crater environment, may wait years for the rain to come to these temporary liquid havens. Some spectacular places, like Hawaii, nurture a bounty of such unique species.

For HIV, every person infected is a new Hawaii—an environmental cul-de-sac in which the intense war with the immune system results in a unique evolutionary result. During each individual immune attack on HIV, many viral changes occur. Protein evolution in the virus accelerates, occurring over a span of years or even months, as the virus rapidly and repeatedly adapts to local immune strategies. In many patients, the first significant evolutionary changes are visible in HIV as soon as ten weeks after the immune system responds to the infection. In just the first year of infection, many protein changes—changes that can have dramatic affects on viral disguise—have been observed to occur on the surface of the HIV coat.

We can display the canvas of HIV evolution for each infected individual by reading the sequences of the viral proteins at different stages of the infection and painting their relationships. Evolutionary stasis would create a blank canvas or faint straight lines like a work of minimalist modern art. Instead, we see something far different:

when viral proteins are examined over the course of the immune wars, we see a startling medusa of rapid evolution.

Reading the War

HIV proteins, like all others, are made of modular components strung together like the stations of a long assembly line. These units are amino acids, and their identity and order determine the function of the proteins they form. Different functions spring from different sequences of amino acids, and in these different strings lies the diversity of HIV action.

Amino acid sequences read like gibberish with an embarrassment of consonants—most of the vowels have flown south for the winter. But with patience, we can read through the inconstant consonants and see the evolutionary shift of disguises that viruses used to evade the immune system of a particular person over a period of years. The first line below presents the amino acid sequence of part of the gp120 protein isolated originally from an HIV patient. Each letter represents an amino acid—the Lego bricks making up the protein. The next three lines report the sequence of the same protein region in the virus taken from the same patient three years later.

```
                                                         Year
CTRPNNNTRKSIHIGPGRAFYTTGEIIGDIRQAHC  1
CTRPNNNTRKSIHIGPGRAFYTTGDIIGDIRQAHC  3
CTRPNNNTRKSIPIGPGRAFYTTGDIIGDIRQAHC  3
CTRPNNNTRKSIPIGPGRAFYTTGQIIGDIRQAHC  3
```

All these lines are different, but seeing the places that have changed is difficult, so let's focus in on the changes by eliminating what has stayed the same:

```
                                               Year
CTRPNNNTRKSIHIGPGRAFYTTGEIIGDIRQAHC  1
.........................D..........  3
...........P.............D..........  3
...........P.............Q..........  3
```

Here, we can easily see that two amino acids have changed in
this tiny fraction of the gp120 protein. All three of the bottom
sequences were present in the three-year-old infection at the same
time, together making up about 40 percent of the total virus population. The top sequence, the original invader, has disappeared
under the first immune-system onslaught.

In year 3 the E has changed to a D or a Q (standing for the
amino acids glutamic acid [E], aspartic acid [D], and glutamine
[Q], respectively), and the first H changed to a P (histidine changed
to proline). All indications are that the P mutation did not provide
a good disguise and that the Q mutation won the replication race.
The evidence for this comes from the sequences observed in year 4,
in which all viruses had that same Q but the P mutation vanished.

```
                                               Year
CTRPNNNTRKSIHIGPGRAFYTTGEIIGDIRQAHC  1
.........................D..........  3
...........P.............D..........  3
...........P.............Q..........  3
.......................V....Q..........  4
..........R........V...EQ...N......  4
```

The last sequence occurred in 46 percent of the viruses surveyed
in year 4, signaling a very successful mutant. The five separate
amino acid changes probably protected this virus from immune

recognition. Of course, the above sequence represents only a tiny fraction of the gp120 gene, and although it sits squarely in the antibody-sensitive V3 loop, amino acids in other parts of the protein also probably changed and played a role in viral success.

Like a pop star, success lasts only briefly, however, and by years 5 and 6, slight variants on the previous themes were being successfully tried.

```
                                                        Year
CTRPNNNTRKSIHIGPGRAFYTTGEIIGDIRQAHC   1
........................D..........   3
...........P............D..........   3
...........P............Q..........   3
.....................V....Q.........   4
..........R........V...EQ...N......   4
..........R.Y......V...EQ...N......   5
......Y...R.G.....SV..AEQ...N......   6
```

At the end of this process, nine of the thirty-five original amino acids have changed and many experimental viruses have come and gone. Some legacies are visible—the Q mutation we noted early on has persisted and all the viruses alive in years 4, 5, and 6 are derived from the single mutant virus that came up with the lucky Q. Three mutations first seen in year 4 also are universal—the R, V, and N changes. Clearly, these detailed views of the virus morphology at the sequence level tell us that viral change within a person proceeds inexorably.

A Forest of Virus

Individual viral evolutionary trajectories like the one above have been recorded many times. Not surprisingly, most evolutionary results differ from individual to individual because viral success derives from random viral mutation, and every person's immune reactions are different. In order to glean from these varied results some of the basic and critical evolutionary patterns, a summary of virus evolution had to be found. Such summaries are in fact common in evolutionary biology, especially since the advent of molecular genetic safaris into the wild kingdom. Using genetic data from just about any critter, plant, or microbe, evolutionary biologists have taken to drawing stick-figure family trees and genealogical diagrams that explain genetic relationships in a clear, simple way.

The rules are as simple as in the game hangman. Sequences are represented by simple symbols. The sequences that are most similar to one another have their symbols connected by short lines. More distant sequences, those that have evolved more mutational differences, are connected by longer lines. There are conventions for how to arrange the lines in a fashion that coddles your intuition about relationships, resulting in a figure like a family tree where similar sequences are grouped together on a single branch.

Trees of HIV evolution represent how particular parts of the exterior proteins like gp120 evolve under pressure from the immune system. When the proteins, which are from sequential isolates taken from individual patients, are compared with one another, we see two kinds of patterns. First, the sequences all tend to start out fairly similar to one another, because usually only a single viral type invades. Within a year, however, the population of viruses has expanded and diversified. On the tree (figure 5.3), we see this represented by a fuzz of short stubby twigs that erupts from near the base. Each of the short twigs leads to a different mutant, usually one with a mutation in the coat or stud proteins. Sequences at the end of different twigs are slightly different from the original virus and from each other. At this stage, the virus pop-

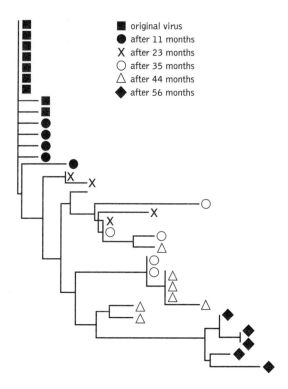

Figure 5.3: A tree showing the evolution of the HIV virus from a single individual over a fifty-six-month period. During this period, the virus evolved continually to evade immune-system attack. The original viruses are shown clustered at the base (left-hand side) of the tree. Genetically different viruses are shown as different symbols.

ulation shows a great deal of variation in these proteins, and that variation fuels the evolutionary stampede.

The same lesson we learned from the amino acid sequences repeats itself here: viruses change rapidly within a person, diversifying into a snake pit of viral types. The tree-based approach lets us compare the results from many individuals and see that this general pattern holds up time and again.

Within two to four years, most HIV infections have changed even more drastically from those present after only twelve months. From among the viral variants visible earlier in the disease, a few have mutated and proliferated into new populations. These were the ones that evaded the first immune attacks and that passed on enough protein differences to allow their viral offspring to hide from the next round of antibodies and killer T cells. They were the callbacks after a molecular audition, and they were the winners. In the virus family trees, these winners are represented by lines leading out of the cluster of original viral sequence symbols.

But like that part of an old family that invested heavily in whale oil instead of petroleum, even a once successful branch eventually dies out when conditions change. The whale-oil barons of the virus world succumb eventually to immune suppression, and their branch does not continue to grow in the virus tree. By four to six years after infection, the viral populations include only those that changed to meet the demands of the newly changed immune system, those that evaded the battery of HIV antibodies and prowling killer T cells looking for previous HIV incarnations.

After years of fighting HIV, the human immune system is like a closet full of clothes you can't wear. Crowded with useless antibodies and killer T cells with no targets, the bloodstream sports only last year's immunological fashions. Having mutated wildly, HIV has won the Russian roulette of immune selection. The loser is the human body, stripped of its defenses against disease by a selfish virus. The virus will win, then die—and take the body with it.

HIV War II: The Drug Battles

After the T cell battle, only manufactured drugs remain as front-line troops against HIV. Substances like the antiviral drug AZT, designed to confound HIV, effectively destroy one type of HIV, but cannot react to HIV's evolutionary barrage. Again, HIV evolves rapidly, randomly and effectively, usually devising solutions to the problems of growing in the toxic drug soup of modern treatments. However, it may not always win. Even HIV has to follow the three principles of evolution in order to survive, and these principles suggest a few ways to succeed against its stacked deck of high mutation rate and fast generation time. Triple-drug therapies are seeing enormous success, but what underlies this breakthrough? Where does the evolutionary engine break down? What are the evolutionary ground rules of our street fight against HIV?

Target I: Reverse Transcriptase

AZT worked for Jeffrey, my wife's patient, allowing him to leave the hospital, returning to his disrupted life. His spirits rose with his white blood cell count, his health bloomed, his yellow Rolls was seen all over town. The drugs were expensive and Jeffrey started running out of money, but AZT, the first AIDS wonder drug, restored hope and health and personal harmony for Jeffrey and thousands of other HIV-infected patients.

But AZT failed. Within a year or so, Jeffrey's white count crashed and his health decayed, plagued by the opportunistic infections that haunt those with AIDS (like Kaposi's sarcoma). Even with his AZT at maximum, AIDS resurged. Across the world, in thousands of AIDS patients, AZT failed and AIDS came back. Why?

RESEARCH UNCOVERS ANTIVIRAL drugs only rarely: viruses make tiny targets. Some effective antivirals work by repressing viral replication or interfering with the construction of new virus coats; nonetheless, other such drugs are difficult to develop because a virus lives through enslaving the metabolic machinery of the host cell, and so we can't easily poison this machinery without poisoning the host. The development of anti-HIV drugs has focused therefore on the few proteins the HIV genome constructs for its own private use. In particular, the reverse transcriptase and protease proteins are vital to HIV life, and both are made from genes carried by the viral genome. Without these proteins, the viral RNA genome couldn't copy itself into DNA, and the exterior coat and stud proteins couldn't be properly processed into new viral globes.

Drugs to inhibit both these proteins have been found. AZT in fact acts as one of the most powerful reverse transcriptase inhibitors, having turned up early in the clinical fight against AIDS. In the human body, reverse transcriptase mistakenly uses AZT when it copies RNA into DNA. One little AZT molecule, stuck into a DNA strand, causes the reverse transcriptase enzyme to gag and stop copying. This thwarts a key part of the virus life cycle, and like putting waterproof bulkheads in the *Titanic*, that should have been the end of worry.

Indeed, AZT did begin working very well, stopping the decline of CD4 helper T cells in most infected patients. But after only a few years, reports of treatment failure began to accumulate. By 1989, virus isolated from patients who had been on AZT for five to twelve months showed increased ability to replicate in the presence of the drug, compared with viruses that had never experienced AZT. Within two years of starting AZT treatment, many patients were no longer protected by drug treatment.

We now know that it takes four mutations in the reverse transcriptase gene of HIV to confer full resistance of the virus to AZT. A reverse transcriptase with all four mutations can operate quite nicely in the presence of AZT because it will not mistakenly use AZT when it copies RNA. Even a single mutation to the right amino acid at just one of these positions makes the virus more resistant to AZT than before. This means that a single mutation can confer a replication advantage, and so, once the first mutation occurs, natural selection begins to operate and the viral progeny of the first mutant spread. Eventually, the second mutation occurs, followed by the third and fourth. At its high rate of mutation, HIV can evolve complete resistance to AZT within two years.

Other inhibitors of reverse transcriptase exist, and they too are effective treatments of HIV—temporarily. The nucleotide mimic dideoxyinosine (ddI) functions like AZT in inhibiting reverse transcriptase and has been used in treatment of patients who have developed resistance to AZT. In many such cases, resistance to ddI evolved within twelve months, often by the substitution of a valine at amino acid position 74 in the reverse transcriptase protein.

Strains of the virus that are resistant to AZT or other drugs transmit easily to other people, and have been observed at alarming frequencies in some cases. A study released in October 1999 showed that more than 25 percent of new HIV infections reported in U.S. military personnel were initially resistant to at least one antiviral drug. Like the transmission of antibiotic-resistant bacteria, the evolution and movement of drug-resistant HIV strains have created a public health landscape vastly more complex, and much more difficult to navigate, than if viral evolution were weak or slow.

Interestingly, selection for ddI resistance seems to reduce AZT resistance, possibly by requiring an amino acid substitution that restores the ability of AZT to bind to the reverse transcriptase protein. A second intriguing finding emphasizes the complicated path of HIV evolution. Dan Kuritzkes at the University of Colorado and his collaborators did a fairly standard study of the emergence of drug resistance in patients taking two different types of reverse transcriptase inhibitors. They found rapid emergence of the viral mutations required to overcome the drug zidovudine (another name for AZT) when it was used alone. But when AZT was taken in conjunction with lamivudine (3TC), emergence of AZT resistance slowed dramatically. Resistance to 3TC evolved rapidly in this and other studies, generally occurring within about twelve weeks after first use. But patients with 3TC-resistant virus have low infection rates and have better clinical courses than patients with uncontrolled HIV.

Mark Wainberg at the McGill AIDS Center in Montreal and coworkers discovered the evolutionary reason for this result, showing how evolution itself had sculpted HIV to be less evolvable. They found that the mutation in the reverse transcriptase genes that confers resistance to 3TC has an unsuspected side effect. It decreases the error rate of the reverse transcriptase, and therefore substantially slows HIV's mutations. The mutant reverse transcriptase has an error rate about three times lower than the original virus. These viruses, forced by natural selection to evolve resistance to one drug, were less able to evolve to resist a second drug. The evolutionary tiger had been tamed—at least a little.

Target II: Proteases

A second weapon in the antiviral war chest, protease inhibitors work by stifling the action of the viral protease and by stopping it from correctly cleaving the virus's coat and stud proteins. As is the case with reverse transcriptase inhibitors, a number of different drugs have been designed to subdue the proteases. But, just as before, HIV quickly evolves resistance.

For example, the development of the protease inhibitor indinavir (IDV) as an antiretroviral drug was accelerated by U.S. regulatory agencies because of its power to reduce HIV growth. Rapid evolution of resistance to this drug also occurred, and within six to nine months, most patients housed viral strains resistant to the drug. No particular set of amino acid mutations were apparent in all resistant mutants, but the overall level of resistance increased along with the number of amino acid mutations seen. The most resistant strains, those capable of growing in the presence of a drug dose thirty times higher than the original killing dose, had four to eight amino acid changes in the HIV protease gene. These evolved strains were also tested against other protease inhibitors (like ritonavir and saquinavir), showing that cross-resistance was common. This shows that evolution of HIV to resist one protease inhibitor generates a viral strain that resists at least some other inhibitors working in similar ways. To date, mutants showing resistance to most protease inhibitors are known, although in some cases the mutations have arisen only in lab experiments. How fast can HIV generate these resistant mutants? For many protease inhibitors, resistant mutants are known before the drug has finished its clinical trials.

Some very careful biochemical work has showed that the resistant proteases do not work as efficiently as the original HIV protein. Biochemists have staged "races" of the HIV proteases that evolved in the presence of protease inhibitors. They were tested against one another and against the proteases found in the original, nonresistant HIV strain to measure how fast the protein completed its chemical reaction. In their experiments, Sergei Gulnik of the Frederick Cancer Research and Development Center in Maryland and colleagues found that the original protease functioned up to thirteen times faster than the mutant ones. They also found, with some exceptions, that proteases with more amino acid substitutions tended to be less efficient than ones with a smaller number of substitutions. These results show that mutations in HIV conferring resistance may also carry a cost in viral replication rate and that the resistant virus, although it has conquered a particular drug, may duplicate itself more slowly than before. So far, however, HIV as a whole does not seem slowed too much by these drops in the efficiency of a single enzyme. Viral mutants with low protease efficiency are still viable, and if these mutants are at an evolutionary disadvantage because of their mutations, that disadvantage appears heartbreakingly slight.

Target III: Multidrug Therapy

About 1996, the big guns were deployed. Independent research on the effects of protease inhibitors and reverse transcriptase inhibitors on HIV levels was combined in a landmark study of the effects on HIV of using *both* types of drugs simultaneously. The results were stunning and enormously encouraging. By using a protease inhibitor plus two different reverse transcriptase inhibitors, physicians could bring HIV levels under control, and CD4 cell counts would then rise. HIV levels decreased by a hundredfold in most patients under this drug regime. CD4 cell counts more than doubled, indicating a rebound in the health of the immune systems of treated patients, although the specific CD4 cells that target HIV—those cell lines that have been destroyed by the HIV infection—never seem to return.

These results have heralded a dramatic shift for people suffering from HIV infection, providing a dose of hope and optimism that has long been needed. Physicians and patients started applying a new word to HIV infections: *chronic*. Long-lasting infections could be survived, and the one thing most AIDS patients in the late 1980s didn't have—time—began to move in the late 1990s at a normal pace for HIV patients.

Cautiously and painstakingly, however, the studies went on. Researchers watched and measured and waited for HIV to evolve a mutant resistant to the triple-drug therapy. All previous victories against HIV had been fleeting. All previous drugs had been quickly evaded. HIV had always won. HIV has always evolved. But this time it didn't.

HIV War III: The Fight Against Evolution

I can think of at least three times that Captain Kirk turned his Horatio Hornblower back on a threat to the *Enterprise*, thinking it sufficiently vanquished, only to see it spring up anew. Nothing satisfies like success, and an all-too-human reaction allows us to declare victory when a terrible and frightening problem appears solved at last. But skepticism about such victory has been the hard lesson in the fight against HIV and AIDS. When the success of triple-drug therapies was first announced, the cautious skeptics hoped like everyone else, but they also waited for the beast to emerge again.

The first triple-drug trials lasted a year, with no sign of resistance. A year later treatment remained effective—a longer course of success than in any previous anti-HIV therapy. Unwilling to wait any longer, several groups of researchers went scouting for signs of evolution. Joseph Wong and colleagues at the University of

California–San Diego School of Medicine found a hidden source of HIV in patients under successful triple-drug treatment. Hidden in a class of cells called peripheral blood mononuclear cells (PBMCs), they found rare, complete copies of HIV. Remember the Rip Van Winkle copies of HIV that enter CD4 cells but then lay dormant (see p. 105)? These rare sleeping viruses can spark new infections and they might be the source of the PBMC copies. But are they really now awakened, original HIV, or had they been just regular viruses replicating unnoticed? Wong reasoned that any rare virus in the blood cells that had been truly dormant wouldn't have evolved because there would have been no cycle of mutation/selection/reproduction. If, however, the rare viruses had been replicating slowly for years, then they were probably becoming resistant, and signs of genetic change in the sequences would appear.

Despite the rarity of HIV in PBMCs, Wong wrung them from the blood cells like water from a damp cloth and looked for signs of evolution. He found none. The sequences were the same as the original. There had been no evolution, and so there had most likely been no replication of these sleeping viruses.

In a very similar study, Diana Finzi and her co-workers at the Johns Hopkins Medical School found dormant HIV in one CD4 cell out of a million, so they are very, very rare. But they don't decline with treatment—they never go away. If they don't go away, they can emerge from their hidden cells anytime and restart the infection. They can start an attack on the immune system just as if the intervening years had melted away like birthdays on a movie star. Only when all the cells infected by HIV have finally died will the body rid itself completely of the virus. Only then can the triple-drug therapy be stopped. Only then would there be a cure.

That's the bad news. But the good news is that the dormant viruses are locked into an evolutionary stasis in which the evolution of resistance seldom occurs. This means that triple-drug therapy may act in exactly the right ways: it drives away the acute HIV infection, and it somehow slows evolution.

There have been failures of triple-drug therapy to control evolution, but these too make sense from an evolutionary perspective. Some failures occur for the same reasons we saw with TB. Interruption of treatment or inadequate dosages can result in a transient release of the infection from drug control. In many cases, the emergence of dormant virus from the PBMCs might spark the new viral explosion. Furthermore, just like in partially controlled TB, drug release can speed the evolution of resistant or partially resistant strains.

In other cases, triple-drug treatment failed when a patient had previously been treated with one of the three drugs singly. Roy Gulick from the New York University School of Medicine and colleagues anticipated this in 1997, when they wrote, "Patients previously treated with lamivudine or a protease inhibitor may not have similar sustained benefits from this three-drug regimen." Prior treatment with a single-drug treatment affords HIV the chance to evolve resistance to the first drug, and when all three drugs are combined, the first one has already been rendered useless.

No Fuel, No Fire

All this still doesn't explain why evolution has been slowed. The seeds of the answer lie in the same strategy used in TB and in other bacterial infections to try to manage evolution. That strategy—to use a drug overkill for which the viral population has no existing adaptations—represents an attack so strong that the evolutionary cycle of viral mutation/selection/reproduction grinds to a halt.

Here, evolution halts for lack of fuel—lack of variation in the ability to grow in the presence of such powerful drug agents. Fuel starvation is further assured because the three drugs all require different and independent mutations to generate any successful resistance. Without viral replication—without a *lot* of viral replication—all three mutations are very unlikely to occur at random in the same virus at the same time.

And so the only viruses left after two years of triple-drug treatment are the dormant ones. They aren't dead because they haven't been active, and they haven't evolved because they haven't been multiplying. They can still do great harm, but at least there are fewer surprises—we know how these old viruses will act.

Jeffrey's Glass Case

Framed in oak and antique brass, Jeffrey's glass case guards a tiny space downstairs. Jeffrey's gift to my wife, after she shepherded his disease through the yo-yo hopes of several years, and a salute to a lost battle with AIDS, it roadmarks the path millions have taken.

Like stones on the Appian Way, carved with the distance to Rome, such monuments of a fight with AIDS scatter across a thousand roads and countless corners of friendships. Monuments of a private war, each chronicles a solitary disappearance, a pact with a devil virus that will not negotiate. Simple things and personal memories, each casts a singular shadow of the human life it remembers.

This legacy of HIV, memorials left behind, documents a human victory of uniqueness expressed in defiance but also in loss. The clever viral chameleon, the blooded street gang, has left a metropolis of monuments while we have learned to fight it. New treatments with the three drugs, like the Three Musketeers, are a thank-god answer to a darkening world. Their swords bring a society hope and raise again the spirit of striving success that the millions of monuments silenced.

Jeffrey might still live if the treatment of evolution existed then. Now we have his monument and others have the chance he lost. And Jeffrey's cabinet is home to our family video collection, a long repository of Christmas gifts and birthday surprises. It's a place of sun and laughter and funny men and spaceships. The glass is now transparent to the loss that brought it here. And on the top shelf sits *The Three Musketeers*, Porthos grinning in a floppy hat like Jeffrey would. It's sitting there, not in reproach, but for all the hope it brings us.

End Game

Despite the success of triple-drug therapies, it's not yet time to relax. True, the evolutionary engine has slowed, but there are still sources of rapid HIV change that could wipe out the substantial gains allowed by triple-drug therapies. Slowing evolution doesn't actually solve the problem—it just gives us a longer time to work out a new answer.

In reality, triple-drug therapies are difficult. They require consistent pill taking—every eight hours for the rest of one's life—with the danger that any substantial drug hiatus will hasten the evolution of resistance. Then we must figure in the cost. Drug treatment charges of $15,000 a year were typical in 1999. This price, too high for most of the people with HIV in the world, practically guarantees that triple-drug combinations will remain the treatment benchmark only in developed countries. Most other people in other parts of the world, where the bulk of HIV infections occur, find this treatment far out of their reach.

The fact that any single drug in a triple-drug combination can be overthrown by HIV evolution in a short time sends a warning note about the future of triple-drug approaches. A virus with resistance to all three drugs is a distinct possibility and once it evolves, it will be very difficult to stop. We may easily say that triple-drug combinations are doing the trick, but the evidence of drug resistance evolving during interrupted triple-drug treatment indicates strongly that the triple-drug success may be short-lived. This worry shouldn't lead to despair; instead, it should reinforce awareness that other HIV strategies are needed. Why do triple-drug therapies work? Because they kill virus *plus* they slow evolution. This strongly suggests that successful strategies should concentrate on slowing disease evolution as much as on killing active or dormant virus.

Many studies of the power of natural selection in wild populations have noted reasons that evolution, despite strong selection, may proceed slowly or erratically. One such reason, conflicting selection, occurs when selection for one trait prevents or even unravels selection for another trait. We have seen such conflicts in HIV interactions with drugs, as when mutations that confer resistance to ddI unravel prior resistance to AZT. Using dual combinations of drugs whose resistance mutations are mutually exclusive may focus the power of evolutionary biology to fight HIV.

Another facet of the evolutionary fight suggests the use of drugs for which resistance mutations confer major disadvantages, such as those that can slow growth rate. An intriguing example was discovered when viruses that have evolved resistance to protease inhibitors were shown to reproduce more slowly than unmutated forms. Perhaps we will discover drugs that require resistance mutations which result in serious reduction in viral growth rates. If so, these drugs may have a longer effectiveness because the evolutionary inevitability of drug resistance might help us engineer a virus with slower growth. As one variant on this theme, evolution of resistance to the drug 3TC dramatically lowers the error rate of reverse transcriptase, thereby slowing the evolution of resistance to other drugs (it may do this by reducing the rate at which genetic mutants are formed). Although this strategy does not slow viral evolution as much as the blitzkrieg of triple-drug therapy, it provides a hopeful glimpse of an evolutionary game plan that might help.

Drugs are evaluated on their potential to kill virus. Fair enough. But the virus is not the only enemy we face. Another foe is the *evolution* of the virus, and few drugs are evaluated on the basis of their ability to kill this process. Furthermore, if drug resistance is inevitable, then by choosing drugs, we are in effect choosing the evolutionary trajectory of the virus. Why not use this opportunity to channel the virus into an evolutionary cul-de-sac and then let loose the pharmaceutical dogs? Granted, we do not know how to do this yet. But while HIV evolves, it thwarts us. And so we must learn how to control evolution in order to survive the evolutionary skills of HIV.

Chapter Six

Poisoning Insects, and What They Can Do About It

W hen Bruce Tabashnik's hunch paid off, he stared aghast, in his office at the University of Hawaii's Department of Entomology, at the first data to ever show that insects could evolve resistance to the world's most potent biological pesticide. From 1986 to 1989, local farm infestations of the tiny diamondback moth had been drenched with the deadly brew called Bt toxin. Much to Bruce's surprise, the moths had evolved resistance to the toxin's liquefying effects, even though no insect anywhere else in the world had successfully evolved this ability. Nancy Cushin, Bruce's technician, wasn't surprised—she had collected the moths and done the careful lab tests. John McHugh who ran the farm, wasn't surprised either; he could tell his sprays were failing. But Bruce knew that when evolution finally conquered the world's best-selling biopesticide, this was big news, and he reached for the phone.

A WALK THROUGH a forest affords a quiet interlude, and even a crowded urban park can blanket the sounds and stresses of concrete commerce. But if you could tune your ears and listen intently, like a new parent straining to hear a noise from the nursery, you would hear an overwhelming sound from all around. From roots deep below ground to the lime green herbs fringing the forest galleries and the dark and serious leaves of the canopy trees, you would hear the sound of insects chewing.

Eating plants is big business, whether in a forest, grassland, or coral reef. And the main plant eaters are not the flashy vertebrates, but instead are small, often hidden, and very very busy. On land, they are the insects, and they chew constantly. They eat roots underground, often as blind grubs tunneling toward the base of a thick-trunked tree. They consume succulent leaves of low, annual plants, and hollow out the seeds made by outraged flowers. They burrow into bark to find the thrumming sap conduits and then plug in to suck their fill. They strip branches of leaves and drop earthward on silk threads to hunt for more. Some even strive for a more delicate eating style, chewing out artistic semicircles from individual leaves as if to carve themselves a natural filigree of lace.

Throughout the salad bar of a forest, lawn, or hedgerow, tiny muscles pull on chitinous levers and operate scraper blades sharper than anything in a hardware store. A toolbox full of insect mouth parts collects plant matter and turns it into more insect body mass.

All this plant eating goes beyond mere cosmetic damage, exacting a much more serious toll in parklands, commercial farms, and orchards. The monotonous expanse of a modern commercial farm, planted with a genetically homogeneous crop selected for high growth and productivity, qualifies as nothing short of insect paradise—a buffet table laid wide and long with a grub's favorite salad. Long growing seasons and constant food lead to generations of insects hatching and chewing, and growing and chewing, and mating and chewing, and laying eggs (and chewing). The resultant explosions of plant-eating insect populations means reduced crop yields, as well as tons of insect pests without any human value—unless someone successfully markets Mrs. Paul's Caterpillar Sticks.

Help Stop the Chewing

Across the United States insect chewing does about $16 billion worth of damage each year. Such crop damage has long plagued farmers, and the attempt to control insect pests has grown along with industrialized farming into a huge business. Humans started poisoning insects soon after inventing large-scale agriculture. In Sumeria, 4500 years ago, insects were killed by dusting plants with elemental sulfur. The Greek historian Xenophon suggests this was not sufficiently effective when he wrote, "Crop protection is in the hands of the gods." Romans were more resolute and added oil to sulfur to create an insect repellent, but they weren't above calling on the gods for help either. This approach was good enough for medieval Europe too—in the twelfth century, panicky Swiss farmers had an infestation of cutworms excommunicated, thinking this would surely do the job.

After the Industrial Revolution, Europeans adopted toxic formulas including largely inorganic chemicals based on arsenic, sulfur, zinc, copper and other deadly metals. Grape vines were sprayed with Paris green, a mixture of copper and arsenic; the same concoction was also used extensively in 1892 to combat the explosive invasion of Boston by the European gypsy moth. Copper and arsenic didn't work in Massachusetts, however, and in 1893, the city switched to spraying residential gardens with a mixture of arsenic and lead.

One of the first powerful insect toxins, an organic mixture called dichloro-diphenyl-trichloroethane (DDT), was developed during World War II and quickly moved into widespread agricultural use. The modern anti-insect arsenal ballooned soon after and now includes organophosphates, carbamates, pyrethroids, chlorinated hydrocarbons, and a new series of bioextracts like the protein toxins from the bacterium *Bacillus thuringiensis.*

These chemicals have been layered heavily and thickly in soils and on leaves. Some of them, like the first organophosphates invented in Nazi Germany for use in nerve gases, kill so powerfully that a hundredth of an ounce will dispatch a bird or rabbit. Nevertheless, in 1996, 14 million pounds of such organophosphates were spread on the corn crop in the sixteen states surveyed by the United States Department of Agriculture (USDA). Another 40 million pounds of insecticide, mostly petroleum distillates and sulfur, fell on the Florida and California orange crops. All in all, 137 million pounds of insecticide were applied in this country in 1996— enough, if directly ingested, to kill every American ten times over.

Corn insecticides—mostly the organophosphate chemicals chlorpyrifos (the active ingredient in Dursban) and Terbufos—are moderately toxic to humans, and very toxic to birds, fish, and other wildlife. Fortunately, and unlike with DDT, they do not accumulate in animals (at least not in live ones) and they degrade within weeks or months to harmless by-products. By contrast DDT was the teacher's pet of the pesticide industry precisely because it persisted for years and kept on killing wherever it had fallen.

Organophosphates attack an insect's nervous systems. Like their sister toxins the carbamate insecticides, these chemicals ooze into insect nerve endings and prevent them from turning off. The insecticide molecules attach themselves to nerve-cell proteins called acetylcholinesterases, thereby preventing digestion of a neural transmitter called acetylcholine. Because acetylcholine remains active in the nerve synapses of a poisoned insect, never being shut off by normal acetylcholinesterase function, once the nerve cell fires, it can't stop. The insect dies a twitching death because of the constant firing of nerves coated with insecticide.

These powerful insecticides have ruled agriculture in the developed world since World War II and have partial responsibility for huge increases in crop yields. Using them can convert a cornfield from an insect paradise to a tenth circle of six-legged hell where faithless nerves betray their owners to a spasming death. But despite this thin layer of insect death in a field of plenty, the ability of a particular pesticide to control a particular insect population usually decays over time. We can poison Mother Nature, but just as surely she will invent alternatives, and soon after heavy use of a pesticide begins, resistance to it starts to evolve.

Loaded with DDT

Fighting just south of Naples was intense in the summer of 1943, and by October 1, Allied forces had swept north into a city left ruined by a vengeful German withdrawal. The summer air in southern Italy had carried smoke of burning factories, but worse was the warning whiff of typhus, carried by body lice from person to person. By winter a full-blown epidemic was gathering steam; and that December, typhus loomed so threateningly that the Allied forces stepped in to control it. They had brought more than artillery to fortify the city and cranes to clear the wrecked port—they arrived armed with DDT.

DDT was uncovered as a potent insecticide by Paul Müller in 1939, who was awarded the 1948 Nobel Prize in Medicine for the discovery. Like ash from Mount Vesuvius, which erupted in March 1944, or Peter Pan dispensing fairy dust, the Allied forces doused the population of Naples with nearly 3 million dustings over seven months. It worked like a miracle, dispatching the typhus-bearing lice and preventing an epidemic that had threatened a million lives.

American troops, casting a penetrating eye on Neapolitan passersby, picked out likely targets and injected such a puff of insecticidal dust under their collars that DDT erupted out their sleeves. By the middle of January 1944, the Allies were inundated with requests for the treatment, and had to set up thirty-one different stations around the city to pump powders. Dr. Ermenegildo Tremblay, now an entomologist at the University of Naples, lived through this intersection of war and plague, "I vividly remember Allied soldiers dusting everyone with a sort of flour we later learned was DDT plus pyrethrum. Dusting operators settled down even in railway stations in order to treat passengers coming in or going out. All this happened in winter, when we were wearing heavy clothes."

Treated houses were marked with acronyms (DDT, MYL, etc.), and the epidemic was officially declared dead in June. But complete victory was short-lived—and only a year later, DDT-resistant insects were reported. By 1946 houseflies in Sweden were resistant, and by 1951, mosquitoes and flies in Italy were resistant not only to DDT but also to a wide range of the new pesticidal chemicals like chlordane, methoxychlor, and heptachlor. Despite the beginnings of resistance, some early successes were beguiling: Egypt used DDT to control mosquito-borne malaria, and the U.S. Congress set up a similar domestic program. From 1947 to 1952, malaria, already on the decline because of extensive drainage programs, was nearly driven from the continental United States.

Such successes spurred greater use of organic pesticides—DDT production was up to 3 million pounds per month by the end of World War II—and led to one of science's major defeats: evolution's victory in the global battle over malaria. Initially, the weapons in this war seemed to favor humans. We finally understood the malarial cycle, in which roving mosquitoes carried a small bloodstream parasite called *Plasmodium*. A quick blood meal would let the parasite slip from the mosquito's syringelike proboscis into an uninfected person's bloodstream, where rapid proliferation would cause fevers, liver damage, anemia, and sometimes death. As long as there were infected people, the local mosquitoes would carry the disease. As long as there were mosquitoes, the disease spread between people like an embarrassing rumor.

In addition to the power of pure knowledge, we also had new chemicals to combat the parasite in the bloodstream and the mosquito in the villages. Quinine, long a remedy known to reduce malarial infection, gave way to new, more powerful drugs like chloroquine after World War II. DDT proved deadly to adult mosquitoes and also to their larvae living in ponds and shallow puddles. However, the DDT was slow to decay, persisting for years and killing with continued vigor whenever insects landed on it. Soon the inexpensive crystalline powders of DDT virtually crusted interior walls in houses all through the tropics.

In 1958, the attack on malaria began worldwide, under a United Nations Global Malaria Eradication Program backed by $110 million earmarked by the U.S. Congress. The funds were slated for a five-year period, for by then, surely malaria would be gone like the world's last car payment. The World Health Organization spread 400,000 tons of DDT around the world and is credited with saving 15 to 25 million lives. But the weapons brought to bear by this ambitious program faltered and eventually failed, all for reasons related to the environmental effects of DDT and the evolution of resistance in both mosquitoes and *Plasmodium*.

Environmentally, DDT was a disaster. Because it persisted and accumulated in larger animals, it triggered a horrifying list of wildlife catastrophes. Widespread fish kills, bird death, devastation of beneficial insect species, and the famous cases of egg shell collapse in predatory birds like eagles and hawks were all caused by the DDT that persisted in the natural food chain after use in pest control. Rachel Carson's 1962 classic *Silent Spring* catapulted DDT into the public consciousness of the United States by showing how indiscriminate pesticide use caused serious ecosystem damage.

Nevertheless, DDT worked in the U.N.'s malaria program at first. Initially, malaria declined steeply in places heavily dosed with DDT. Sri Lanka, with a million cases of malaria in 1955, reported less than two dozen in 1961. That same year, reports showed the existence of resistant strains of both mosquitoes and *Plasmodium*. The parasites had evolved resistance to the chloroquine drugs while the mosquitoes had evolved resistance to DDT, thus bringing the attack on malaria to a standstill.

The five-year time frame for malaria eradication ran out without a victory. Why didn't the program continue? Because the program's architects were racing evolution, and they thought if the eradication program progressed too slowly, then "resistance is almost certain to appear and eradication will become economically impossible."

So the race went not to the hare but to the tortoise, a tortoise fleeter by far than anyone had suspected. By 1972, the World Health Organization, having spent $1 billion on malaria eradication, declared the program dead. The developed world turned its back, leaving developing nations with an exacerbated health problem. After partial "eradication," the populations of these countries were poorer in immunity to malaria than before, and were plagued by new strains and new mosquitoes that are very hard to kill.

The Resistance Movement

Like many adults, Richard J. Pollack of Harvard's School of Public Health may tousle the hair of preschoolers as a friendly hello, but for him it's all business. Pollack and his colleagues study head lice among children, and recently compared lice found on American and Malaysian heads. In the United States, lice are an inevitable, if unpleasant, accompaniment to day care and elementary school, treated by dousing with medicated shampoos. In many places, a parent can't bring an infested child back to day care until a week of washing with insecticidal suds is completed. However, Pollack found the U.S. lice much less responsive to standard treatment than head lice from children in Sabah, Malaysia. Malaysian lice, he discovered, had no history of treatment with medicated shampoos and showed no resistance. But in America, where medicated treatments are the norm, lice have evolved to be much more difficult to eradicate.

The first studied case of insect resistance focused on the San Jose citrus scale insect in California that evolved a tolerance to sulfur treatments in 1908. By 1967, 224 species were known to be resistant to insecticides. By 1986—a year in which Environmental Protection Agency records show that 188 million pounds of insecticide were applied in the United States—there were 447 known resistant insects; and by the late 1990s that figure topped 500. Resistant populations range from mosquitoes to potato beetles and gypsy moths, and they occur in every state and most countries, damaging crops and increasing exposure to diseases. Virtually every insecticide has been overcome by at least one insect species.

Tobacco budworm, a serious pest on cotton, evolved resistance in the 1960s to virtually all available pesticides in an area of northeastern Mexico and Texas's lower Rio Grande Valley. By the

1970s, approximately 700,000 acres of cotton fields lay abandoned because this moth's resistance was so high that the crop simply could not grow. Other studies show that over time insecticide resistance requires increased chemical use; in California cotton fields, 74 percent more organophosphates were required in 1981 than in 1979, even though net reduction in insect population was the same.

In 1904, 10 percent of agricultural production in the United States was lost to insect pests. In 1986, despite over 100 million pounds of insecticides, insect losses stood at 13 percent. This value would probably be lower except for dramatic changes in U.S. agriculture that favor insect pests: huge crop monocultures—miles of cotton, corn, or soybeans—and a shift to more insect-sensitive crops. But insect resistance also deserves a great fraction of the blame for the rate of pest damage. Today, insecticide resistance in the United States increases pesticide use by about 10 percent, through need for increased dosages and need to apply more than one chemical. Thus, resistance costs U.S. farmers about $1.6 billion a year.

MOSQUITOES HAVE BEEN targeted by some of the world's most intense pesticide programs and, consequently, have come up with more strategies for thwarting control than a convention of tax lawyers. In 1989, 114 different species of mosquitoes resisted at least one insecticide. Many had developed multiple resistance, especially to the powerful organophosphate neurotoxins. Resistance comes from a number of different cellular mechanisms that prevent the attachment of the insecticide to the insect's nerves. In some cases, mutations in the acetylcholinesterase protein rescue it from being attacked by the insecticide. In other cases, another enzyme—actually another esterase—attaches to the insecticide before it gets to the nerve ending, effectively blocking insecticidal action. When there is a lot of insecticide around, this strategy works like shoveling your driveway during a heavy snowstorm— it's successful only if you have a lot of shovels. But mosquitoes make a lot of chemical shovels (the esterase enzymes) to scrape up the insecticidal snow because, not so long ago, a mosquito invented a way to increase the production of its chemical shovel many times over.

Sometime, somewhere, while heavy doses of organophosphate insecticides were first being used, the esterase gene was duplicated many times inside the cells of a mosquito. Multiple copies of a gene, residing in a cell like a bank of mechanical fabricators on an assembly line, produce much more protein than a single gene. This overproduction increases the amount of the protective esterase, and allows detoxification of even massive doses of the insecticide. The new strategy was extremely successful, and from its first recorded appearance in 1986, spread around the world like a good internet joke.

How did this mutant trait appear so suddenly and so omnipresently? Two possibilities exist—that the mutation appeared several times independently or that it spread like lightning from one place. Careful genetic work has examined small snippets of DNA from resistant individuals of the mosquito *Culex pipiens* from around the world. This work tells us that the DNA surrounding the duplicated genes is identical in mosquitoes from California, Pakistan, Texas, and Egypt. Because we expect such duplications rarely to happen independently in four different parts of the world, it seems that just one single mutation—a drastic one—caused esterase duplication in this species. Live adult mos-

quitoes fly internationally, hidden on airplanes, and easily leap
across continents. So amplified esterase genes soon embarked on a
global tour, leaving progeny everywhere they stopped.

The Price of Success

But evolutionary changes such as those observed in the mosquito,
like bargains with the devil, carry a cost. A great deal of raw materials are needed to make overproduced esterase, and insects that
needlessly make so much of the extra protein are selected against.
Field studies show that, in the absence of insecticides, mosquitoes
without the overproduction of esterases grow faster, survive
longer, and reproduce better than the resistant strains, and that
areas without heavy pesticide use have fewer mosquitoes with the
amplified esterase genes. Natural selection against resistant individuals reduces their frequency in subsequent generations—as long
as the insecticides remain absent.

But evolution can run subtly, driven by selection to reduce the
costs of insecticide resistance—and mosquitoes that pay the devil a
discount price can thrive. For example, most mosquitoes overex-
press, or make too much, esterase and wastefully spread it throughout their bodies into tissues where insecticides have no effect.
Other mosquitoes produce the protective protein only in their nerve
cells, the tissue that needs protection most. This second group outstrips the first in growth rate, survival, and pesticide tolerance.

Other mosquito strains also produce esterases in the intestine
and the cuticle, where the insecticides enter the insect body. These
may act like molecular scrubbers, tiny organic robots attaching to
insecticide molecules, deactivating them before they get close
enough to nerve cells to do any damage. This difference in the position in the body from which the esterase launches its defense is an

ongoing experiment in the evolution of increased pesticide tolerance, and one that hasn't spread completely around the globe yet.

They Don't Check Out

The urban cockroach, sometimes called the German cockroach,
has a detailed record of pesticide resistance, evolving so quickly
that in the 1980s and 1990s they seemed on the verge of overwhelming the insecticide industry. Why aren't we up to our knees
in cockroaches by now? The answer lies at least partially in the
success of nonchemical methods of control. Cockroach motels
attract animals to their death, and reduce populations so much
that many fewer resistant individuals occur by chance. In a population of 1 million roaches, there's likely to be one resistant to any
particular pesticide, and so if we exclusively use chemical methods
to reduce the problem, then resistance evolves quickly. But a population of 100,000 roaches has only one tenth the chance of including a resistant individual; as a result, evolution is less likely.
Approaches that rely on more than pesticides to control insects are
called integrated pest management (IPM), and have become the
hallmark of effective approaches to managing both insect pests
and their evolution. Like the direct observation therapy used to
control the evolution of TB, IPM works because it simultaneously
attacks a specific infestation and addresses the larger problem of
slowing the evolution of resistance.

The Transgenic Revolution

Not all insecticides are christened in a percolating chemical vat of
murderous molecules. Some with the widest use flow from natural

products of organisms that long ago invented ways to kill insects for their own purposes. We have co-opted these organisms, using their toxic products to produce a killing paste that works better on specific insects and kills fewer of our wildlife friends than the organophosphate alternatives. The most popular "natural" pesticide is the Bt toxin mentioned earlier; made from the bacterium *Bacillus thuringiensis*, it kills specific caterpillar pests on demand and constitutes up to 90 percent of the biopesticide market.

Dark mornings rule in wintertime Seattle, where a slow battle rages between the distant mountains and the head-high cloud layer to see which will be the last to let the foggy sun in. When I was in graduate school there, these dim transitions gave way to another contest—between little caterpillars that ate trees and the determination of the Emerald City. These small but hungry caterpillars died only at the hands of Bt toxin.

At the time, the gypsy moth had denuded square miles of deep-needled conifer forest throughout the Northwest. Despite major efforts within the forests themselves to stunt the gypsy moth population, the suburbs and city parks of Seattle remained a pest sanctuary. Eventually, Seattle hired a flock of thundering helicopters to coat the infested neighborhoods with buckets of Bt toxin.

Shortly after dawn on the appointed day, the helicopters swept the clouds aside with hissing sprayers and thundering rotors. My housemates and I stood aghast as the very air was ripped apart by the roar of rotating blades. The flying nozzles sprayed cars, houses, and streets and the occasional tree with little white dots of Bt toxin, turning the city into an explosion of a baby powder factory. The bugs died, and we washed our cars, scrubbed off our bikes, and hoped we'd never have to hear that concussion-blade noise again in the still Seattle air. We went back to work and to school and the normal university noises. Gypsy moths were history.

Beautiful Crystals

But they weren't. Seattle sprayed again, in 1999, to battle the gypsy moth. Vancouver had already done so in 1991. All this highlighted the deep impact of Bt toxin, one of the most pivotal industrial evolutionary experiments of our time. Throughout the world, humans grow vats of the *Bacillus thuringiensis* bacteria and use their toxins to eliminate caterpillars of corn borers, diamondback moths, cabbage moths, gypsy moths, the western spruce budworm, and a Pandora's box of other pests. New designer varieties of the same bacterial toxins, like perfumes of deadly intent, kill mosquito larvae, flies, or even the Colorado potato beetle.

Bt toxin (sometimes abbreviated to just Bt) has a load of advantages. Because it is a protein, and one produced by a biological agent, crops sprayed with any variety of Bt are still considered organic. Even while Bt toxins are deadly to their targets, they do less harm to the rest of Earth's species than any of the chemically based insecticides that grace the pressurized tanks of modern agriculture. This is because Bt toxin kills an insect larva from within, depending on an attack strategy tuned to particular pests.

Bt is produced by bacteria to penetrate and kill a new host to invade and consume. When environmental conditions deteriorate, *Bacillus thuringiensis* bacteria make tough spores that can scatter into the winds and wait for a new insect host to come along. A delicate crystalline lattice of inert precursor toxins coats the spores—nontoxic until activated inside an insect gut. The spores and their coiffure of soon-to-be toxins lie scattered on leaves and stems, where their fate is to end up eaten by innocent herbivorous insects.

Once inside an insect, the crystalline Bt toxin dissolves in the gut, remaining inert at first. Bt proteins are formed in a long chain:

the active toxin occupies one end and a throwaway protein takes up the other. Only when a normal insect digestive enzyme cuts the two parts of the chain apart is the active toxin released. It quickly attaches to another insect protein on the wall of the gut, punching a tiny hole. Soon, thousands of holes—perhaps millions—open up in the insect's gut, and like a rush-hour subway train disgorging at the station, the gut contents spill into the body cavity of the doomed insect. Bacterial spores ride this wave of fluids, and the insect dies in its digestive juices, providing food for the next generations of *Bacillus thuringiensis* that are by now hatching from the tough spores.

Humans stumbled across these nasty table manners in 1901 by discovering *Bacillus thuringiensis* in the body of an infected silkworm in Japan. In 1911, it surfaced again, in Europe this time, but not until 1958 was the Bt toxin produced by this bacteria produced commercially as an insecticide. Bt provides strong protection of field vegetables, corn, and cotton crops, as well as the last line of defense against deforestation by invading gypsy moths. New boutique formulations of Bt toxin, like *Bt israelensis*, have been used extensively to kill disease-carrying blackflies.

By far the best feature of Bt toxins has always been that resistance to it evolves slowly. The first insects resistant to DDT emerged just three years after its use to combat typhus-bearing lice in Naples, even before Paul Müller won his Nobel Prize, and the general pattern of resistance appearing within a few years of a chemical's commercial development has been confirmed over and over. Yet, resistance to Bt proteins remained unknown three decades after their introduction, always killing, always working on the targeted populations.

Then, in 1989, that changed.

The Moths of Sumida

Sumida Farm, the incongruously green neighbor of the biggest shopping center in Hawaii, perches between the steep volcanic slopes of Aeia Heights and the blue-gray shallows of Pearl Harbor. Sharing its spring-fed valley with the crowding push of Pearlridge Mall and the recent crop of fast food chains and muffler franchises, the bursting green of Sumida Farm breathes a moist breath of agriculture out into the commercial jungle.

John McHugh joined Sumida Farm with a degree in horticulture and a farmer's ambition. He applied his knowledge and determination in equal degrees to turn watercress into gold, adding an experimentalist's glee to the Sumida family business, which had been growing since 1928. An aquatic dynamo, watercress grows like a weed in wet, rich soils bathed in clear running springs, and serves as a delicate green layer supporting the creative Asian cuisine of Hawaii.

The brilliant green leaves have many fans, none more discerning of a fine, fresh crop than the diamondback moth. The caterpillars savage the leaves and stems, damaging even more than they consume, leaving a crop badly burrowed and commercially useless. And imagine finding a caterpillar parboiled between the watercress fronds on your platter of ginger-sauce *ono* with pineapple-mango chutney.

For years, chemical insecticides were the solution to the diamondback problem—a coating of chemicals that could be washed off before the perfect fronds were used. In the early 1980s, most watercress farmers in Hawaii followed the same routine: spray every week, month after month, during the nonstop growing season.

In 1982, this routine collapsed as the moths won round after round of the evolutionary battle. Insecticides failed that year—the moths had evolved resistance to them all—and farmers watched their crops wither to a brown watery scum. In desperation, they forgot USDA regulations and sprayed the fields with banned organophosphates—anything to stop the onslaught. The surreptitious spraying rescued the crop, but was done in a panic, without full safety precautions. One farmer collapsed in his fields from organophosphate toxicity. Five times, the state of Hawaii found insecticide residues on marketed watercress that had been sprayed too close to harvest. Five times farms throughout the state were shut down.

Seldom had the diamondback moth been so consistently victorious. But Sumida Farm seemed a little less hard-hit than others. "The whole farm was a wreck," John McHugh told me with a wry grin, "except those patches where I had set up my sprinklers."

We were standing in the middle of the farm, the sprinklers erupting around us in a choreographed pattern controlled by unseen valves. Cattle egrets patrolled the spring-fed rivulets, feasting on tadpoles and mosquito fish. The midafternoon sun hit hard, and we stepped into the shade of a lone papaya tree. John had discovered that spraying the fields with regular geysers of spring water not only kept the leaves cool but knocked down the flying moths, and reduced egg laying. "I started watering the crop so it would grow better in the summer heat," he explained. "But it was the only thing that kept the moths away in 1982." John added a final innovation, Bt applied liberally to emerald leaves, and Sumida Farm thrived, its water sprinklers punctuating the Hawaiian evenings.

By 1986, the story switches to Bruce Tabashnik's laboratory at the Department of Entomology at the University of Hawaii. Bruce was researching why diamondback moths on the islands had evolved such broad resistance, especially to the pyrethroid insecticides, becoming a scourge not only for watercress but for cabbage and other crops. Tests for pyrethroid resistance were standard fare in Bruce's lab, but John insisted that something was different about his moths, arguing that they should be tested for resistance to Bt toxin. Bruce agreed, even though Bt resistance had never been found in the field in any insect before.

"We ran the first tests in 1986 and indeed found reduced susceptibility in larvae derived from John's watercress relative to our standard lab strain," Bruce reports.

But a moth in the ointment was that other caterpillars in Hawaii, not sprayed extensively with Bt insecticides, showed no difference to the Sumida Farm critters. If resistance had evolved, it was slight, and not centered at Sumida. Emboldened, John sprayed Bt with a vengeance, but over the next three years saw his crop decline in weight and value. When he went back to Bruce in 1989, they found an astonishing evolutionary change.

"Populations that were heavily sprayed had increased resistance relative to 1986," Bruce remembers, "Voilà! Direct evidence for rapid evolution."

McHugh is just as blunt: "I got away with three applications of Bt, but by the fourth, the moths were completely resistant."

One of agriculture's safest ciphers had been decoded by evolution's persistence. The word flashed across the agricultural world that Bt toxin could no longer qualify as the unbeatable insecticide. How had the moths of Sumida done it? Why were they different? And could the evolutionary damage be contained?

But Would It Have Worked on Mothra?

Insects in laboratory populations respond to selection for Bt resistance, much as artificial selection in other insects can lead to hairy flies or curly wings (see chapter 2). If laboratory populations will evolve, why not populations in the wild? The question was of mild academic concern until Tabashnik showed the honeymoon was over and that wild populations of the diamondback moth could feast on watercress crops dusted heavily with Bt toxin. But even more important, Bt toxins were adopted by a number of biotech companies as the best way to engineer an insecticide into the genome of crop plants. If Bt was resistible after all, the genetically modified crops might as well be expensive caterpillar salad, and a great deal of fancy bioengineering would be at serious risk.

In fact, the first trials of genetically modified crops did not fare well. In 1996, 1.8 million acres were planted with cotton that had a Bt toxin gene inserted. Molecular geneticists had removed the gene from the bacteria, and spliced it into the chromosomes of cotton. Now, instead of repeatedly spraying Bt toxin on a crop to ensure plant-eating insects got a good dose, the bioengineered cotton plants contained the toxin within every leaf and stem. The cotton made a particular protein named Cry1Ac, which is toxic to some of the common cotton-eating insects. But other insects nevertheless overwhelmed the Bt cotton, and farmers who had already paid high prices for cotton seeds with the Bt label had to spray with conventional insecticides.

Seed companies shot themselves in the foot by claiming incorrectly that Bt cotton would quell all important cotton pests. But high temperatures and the close proximity of corn crops that harbor relatively resistant bollworms increased the number of insects that could find the Bt cotton crop. These insects are well-known to be resistant to the Cry1Ac protein and, without the normal toxic cocktail produced by *B. thuringiensis*, they were largely immune to the cotton's new genes. For the bollworm, for example, more than 10 percent of natural-born caterpillars could resist the low Bt toxin levels that the 1996 cotton crop produced. In addition, one out of 500 individuals in another major pest population, the tobacco budworm *Heliothis virescens*, carried a mutant gene that conferred resistance. Such genetic variation for Bt resistance shovels fuel into the evolutionary firestorm, requiring current cotton planters to warily search for Bt resistance in the field.

SO FAR, THE mystery only deepens. Bt toxins worked well for three decades, becoming the most common biological pesticides in the world, and only the mild-mannered diamondback moth eventually overcame its toxic effects. Yet the first foray into biologically engineered Bt toxins in crop plants fell short of a success. Even more odd, laboratory populations of insects have been shown to easily evolve Bt toxin resistance, and field populations of major pests possessed Bt resistance genes even before the Bt cotton crop was planted.

Why doesn't Bt resistance evolve in most field populations? What makes field and lab experiments so different from each other? The answer may lie in the interlocking complexity of biological evolution, which has produced a full complement of Bt toxins tuned to different caterpillar prey. The array of Bt toxins is like the sharp diversity of fishing lures on display at any tackle shop. You got your Buzz Bombs and Crazy Crawlers, designed to attract different kinds of fish and fishermen. Just like choosing today's lure may need a careful pondering, picking the right Bt toxin requires understanding of what each one is good for.

A Bt mixture consists of several different proteins in a cocktail produced by the bacterium. The strain *Bt kurstaki* HD-1 has toxic proteins Cry1Aa, Cry1Ab, Cry1Ac, Cry2A, and Cry2B, and some insects are more sensitive to these than others. Moreover, most sprays also contain the waiting spores of living bacteria, hungry for food. Once dumped on pests, they hatch out, and like tiny piranhas gobbling caterpillar innards, they crown the killing power of the Bt treatment. The naturally produced Bt spray contains so many different pesticides that few insects can resist them all. Like the multipronged attack on tuberculosis in which many different antibiotics simultaneously overwhelm the bacterial population (see chapter 4), Bt overkills naturally, with more deadly concoctions than a bar in Key West.

Whether a Bt mixture works on a pest (and whether resistance easily evolves) depends in part on the strength of the toxins on that pest: more effective killers slow the evolution of resistance. One careful lab study tested evolution due to a single toxin, comparing results with a mixture of four different toxins; this study showed that single-toxin evolution was rapid but that lab insects exposed to all four toxins at once, or to toxins that were particularly effective, evolved slowly or not at all.

But complex mixtures by themselves are not invincible: other studies showed that resistance in the lab could evolve quickly even when insects were treated with entire Bt cocktails. So why no evolution in the field? New evidence traces sluggish evolution to the effect of sunlight on field experiments; sunlight is of course missing in the lab. Sprayed on the surfaces of leaves, Bt toxins decay quickly in the sun to harmless dust, rendering any application a quick, transient poison. Selection is powerful for a brief moment, but soon ends as the toxins decay, leaving the treated field ripe for reinvasion by pests from elsewhere. Under these circumstances, evolution is disjointed. High mortality leaves extremely few insects to breed, and the next generation of pests probably springs from adults that waft into the field from next door—without genes for Bt resistance. Only if the Bt applications persist—like when John McHugh sprayed over and over in Hawaii, or when single-minded lab experimenters consistently dose their caterpillar cultures under dim lighting—does evolution speed up.

The Difference Between Physical and Biological Engineering

The lesson from this evolutionary realization? That biological engineers have a serious unanticipated problem. They have engineered a poison that is not destroyed by sunlight. Instead, the leaves produce it inside, sheltered from UV damage by the leaf tissues, where it can always kill, as if there were a million intracellular John McHughs applying Bt toxin to the insides of every cotton leaf and corn blade.

The poison does not fade, but neither is it designed to be a massive overdose, an overkill strategy that completely eliminates a pest, and it shows none of the biological complexity needed to quell all the pests present in a field. The biological complexity of multiple toxin genes, plus the added time bomb of the waiting spores and the inactivation of the poison by sunlight, may constitute the only recipe to prevent wholesale Bt resistance. Bioengineers have taken the easy way out in this respect, by inserting only a single Bt gene into crop plants. To be fair, inserting a single gene would be the place to start—and even this isn't easy. But such a simple start to biological engineering may spell evolutionary disaster if the engineering effort stops there. Like offering a tuberculosis patient a single drug at a time, dosing insect pests with a constant level of a single toxin gene product may be the best way of speeding the evolution of resistance. Once resistance to that single toxin spreads, more complex toxin cocktails may fail much more readily.

Here's my formula to speed up the resistance of insects to Bt toxin. First, add a single gene to a crop plant, and have that gene produce such a mild dose of Bt that a good fraction of insects (say

1 to 10 percent) can survive. This describes the 1996 Bt cotton crop pretty well. Next, when resistance appears, add another gene to the crop and increase the dose of the first gene (expectations are that Monsanto will introduce a two-toxin line of crops within a year or two). Resistance to the first gene will increase, and insect populations already partially resistant will have the best chance of evolving resistance to the second gene. Repeat this procedure until natural mixtures of four or five Bt toxins have been gradually engineered into crop plants. But by then we will have walked the insects slowly to evolution of full resistance against the complete mixture—a walk they may never have taken had they been faced with the full toxin battery at the beginning. It turns out that by chance a single mutation in a single gene conferred resistance to all Bt toxins in diamondback moths; however, other insects, without this ability, will have to be carefully taught to overcome Bt. So far we are good teachers.

Suppose rivers could evolve resistance to bridges, somehow developing methods to prevent their being crossed by wood or stone or iron? How would engineers have changed their bridge designs over the millennia? Would pillared roadways slowly have changed to viaducts supported by Roman arches? Then finally to boxy cantilever and finally the pinnacle of the suspension bridges of the Industrial Revolution? Or would this slow progression merely have taught streams just exactly the mechanisms they needed to keep from being spanned by human industry? This is a fanciful notion, one so divorced from the reality of solid engineering that we might struggle to conceive a world saddled with this kind of problem. Physical engineering, existing in an evolutionary vacuum, has the luxury of ignoring the selection that engineering solutions can exert on biological targets.

But biological engineers have generated these problems ever

since they first began to alter species to suit human needs, and they have found that other species soon evolve to thwart human plans. Such engineers of living things face the evolutionary dangers of incremental solutions to complex problems. They cannot ignore the evolutionary stage on which their technology projects perform and cannot discount the power of evolution to generate rapid change.

To be sure, evolution of Bt resistance now worries even economists and mutual fund managers. Nonprofit groups like the Rockefeller Foundation—perhaps with an eye toward longer-term benefits—are reported to be developing three-toxin transgenetic plants. Stock prices of the seed company that produced the 1996 Bt cotton crop fell 18.5 percent in a single day when news broke that parts of the crop were overwhelmed by pests. The pests and the stockbrokers sent a message that day: evolution exerts powerful economic forces that have become a critical part of the biotechnology industry.

Interestingly, the preferred solutions to reducing the rate of Bt resistance seem very different than the solutions proposed to limit the evolution of antibiotic resistance. For example, the Bt world ignores the direct observation therapy solution so successfully used to reduce TB resistance. The TB strategy focuses on local eradication where treatment overkill eliminates the disease in a single person without giving it a chance to evolve. Such a strategy does not seem practical for individual agricultural fields, partly because treatment overkill has environmental consequences.

Deadly Dust

One reason that treatment overkill works poorly in this case is an unexpected effect of high levels of Bt toxin production on nonpest insects. Bt toxin genes often turn on in all plant parts: leaves and stems, fruits and flowers. But one plant tissue—one that in fact should not be protected—does not need to be guarded by Bt toxin because this part, pollen, departs the plant to float away on the wind. Nevertheless, pollen can create an environmental cloud of engineered gene products by carrying the toxin away from the carefully tended fields of bioengineered plants and transmitting the engineered genes across the landscape.

A pollen grain carries a plant father's genetic contribution to the next generation in a finely sculptured packet of proteins and waiting cells. Those stingy flowers that pass their pollen on a bee's whim prudently parcel out their bounty, and produce just enough pollen to coat a bee or two. But some plants, especially grain crops like corn, pass their genes by spreading pollen into the air, and these flowers produce a prodigious supply of pollen grains—most destined to float as dust in the wind. Pollen makes for a characteristic allergic reaction when it mistakes your nose for a flower.

Currently, U.S. farmers plant millions of acres of corn engineered to produce Bt toxins. This golden blanket, like corn everywhere, releases pollen in dusty floods. Pollen carries along in the breeze, and finds its way everywhere we live, coating cars, homes, and other plants with a particulate lace. Most Bt pollen carries the toxin along with it, coating other plants with small packets of Bt protein that insects readily consume.

Recently, a simple experiment has demonstrated the potential importance of this unanticipated vector of pollen poison. Monarch butterfly caterpillars eat only milkweeds, which grow throughout North America, especially at the edges of crop fields. Pollen from adjacent cornfields frequently dusts these plants during the height of the summer—just when the monarchs have completed their northward migration and have begun laying tiny cream-colored eggs. When milkweeds dusted with pollen from a Bt corn crop were fed to monarch hatchlings, the caterpillars grew more slowly than usual, and nearly half of them died during the four-day experiment. By contrast, none died from eating normal milkweed leaves, and none died from eating leaves dusted with corn pollen that did not contain Bt genes. So, clearly Bt pollen kills caterpillars of non-pest species and this killing potential can move beyond the crop field into surrounding ecosystems.

Several critical lessons sprout from this experiment. First, although control with Bt toxins may not affect birds, fish, or mammals, other insects can be killed when Bt escapes the farm. The weather and the industriousness of pollinators export Bt toxin from a particular cornfield, and we must decide what risk this poses for ecosystems outside farms. Second, this simple experiment came as a rude shock to a $1 billion agritech industry, which had no comforting response when the monarch results hit the newsstands. Despite the fact the monarch results never really showed that levels of Bt pollen were high enough in the field to kill benign insects (and in fact subsequent studies on the less sensitive swallowtail butterfly showed no effect), the government of Austria banned planting of Bt corn, and the European Union—skeptics of biotechnology's broad claim of harmlessness—reconsidered approval of the crop technology. A scientific meeting called in the summer of 1999 to review these results, plus follow-up studies conducted by scientific, industry, and government groups, found evidence supporting the export of toxin by pollen, but insisted no widespread environmental damage results.

U.S. regulatory agencies consider biological engineering no different than other forms of selective breeding. FDA approval of genetically engineered foods has been hastened by considering all gene modifications as "generally recognized as safe," thus avoiding the need to exhaustively study product safety. But when the monarchs died, some of the unforeseen consequences of this massive and rapidly engineered evolutionary change emerged, and the ecological complexity of our evolutionary tinkering began to come to light. It is not unfair to expect multinational corporations to understand about pollen, especially when they make it kill.

Pest Refuges and Human Altruism

Adding the Bt gene to a plant will not produce a pesticide by itself— the gene must also function in the plant to produce the toxin. This means that in addition to the gene itself, genetic engineers add a welter of other control sequences, including a bit of DNA called a promoter, that regulate when, where, and how much of the toxic protein a plant produces. We may desire a high level of toxin production in plant tissues for overkill control of pest populations, in much the same way that taking a large dose of antibiotics can control a bacterial infection. But this overproduction/overkill strategy may also make pollen more toxic than before.

Other reasons exist why the treatment overkill strategy, though it works for human diseases, fails to slow insect Bt resistance. Insects can reinvade a field more easily than diseases can reinvade a person. Disease treatment seeks to rid a person completely of the pathogen because even a small residual pathogen population can reestablish a disease. But the economic costs of a mild insect problem may be easily overlooked, and we generally ignore rare pests.

Such a casual attitude may disappear when the pests can overcome Bt engineered into crops at great cost. What other strategies, then, will slow the spread of Bt resistance, extend the commercial lifetime of the expensive genetically engineered crop plants, and allow overall reductions in chemical insecticide use?

In 1999, Monsanto locked farmers buying their Bt-engineered seed into a theoretically sound program of evolution reduction called refuge planting. In this strategy, the farmer plants a field unprotected by Bt near the Bt-protected crops. Insects thrive in these so-called refuge fields, and do not evolve Bt resistance. When the adult insects breed, we expect they'll mix between the unprotected and protected fields, thereby diluting the resistance genes that have been increasing in frequency in the Bt-protected field. Experimental field trials and computer simulations that mimic this mix of crop treatments show that this strategy can work to slow the evolution of Bt resistance for two reasons.

First, many of the genes that protect insects from Bt are partially recessive, that is, an individual must inherit Bt-resistant alleles from both parents in order to receive the full benefit of Bt resistance. Because of this, diluting the Bt-resistant alleles with alleles from refuge populations works: in this case, most adults with a single resistance allele will mate with insects without any resistance allele, and the resulting progeny will have little resistance ability. Only rarely will two resistant individuals find one another in the crowds of swarming insects—like Humphrey Bogart trying to find Ingrid Bergman in the crowded Paris train station, although in *Casablanca* a very different type of Resistance is trying to evolve.

Second, insects with Bt resistance often suffer a disadvantage in competition with members of the same species without Bt resistance. Perhaps resistant insects might have lower levels of important digestive enzymes or may avoid some good feeding locations like leaf surfaces. No matter what the mechanism, if Bt resistance carries with it a certain cost, then selection will act *against* it in refuge populations. So, farmers and Monsanto have evolution working against them in the Bt-engineered fields, but working for them in the refuge fields. The balance might favor the farmer, if selection acts strongly against Bt resistance in the refuge, and a large number of the insects mature in the refuge rather than in the Bt field.

Although a useful evolutionary standoff might lurk within this arrangement of refuge plantings, an interesting economic tension has developed between the farmers and Monsanto. Perhaps, as hoped, there will a long-term benefit derived from planting refuges—perhaps the evolution of Bt resistance will slow. This could increase a farm's productivity in the long term and repay Monsanto's considerable development costs for the Bt project. But the farmer pays an immediate price—a short-term loss of field productivity. For the refuge strategy to successfully dilute evolution, each refuge must produce a bucketload of insect pests, and the refuge field will therefore be an economic write-off paid for by the farmer. From the farmer's standpoint, it pays to cheat—to not plant the refuge and thus avoid the price of lowered productivity. Even if the revenue from the Bt field offsets loss of product in the refuge field, a farmer gains in the short term by secretly planting the refuge field with Bt seeds or heavily spraying it. This reduces long-term gain for the agricultural industry but increases short-term profit for the farmer.

Resource managers all too easily recognize this situation, termed the "tragedy of the commons" by ecologist and philosopher Garret Hardin in 1968. The "tragedy" is that individuals gain the most when they take as much as possible from a commonly used resource pool, even though such rampant overexploitation might deliver disaster in the long term. Hardin applied this concept to use of common resources like fish stocks or clean water, but it extends easily to a situation in which the common resource consists of time until evolution of resistance. By cheating, a farmer can slightly decrease this time, but at an advantage to him- or herself.

What will happen when individual gain competes against a societal need to slow evolution of Bt resistance? Will Monsanto's strict sales agreement police farmers so well that they always plant refuges? Will "improvement" of the biologically engineered plants fix the problem of rapid field evolution of resistance? Or will incremental engineering breed such careful selection on insects that most become resistant even to natural Bt toxins, thus eliminating one of our best biological insecticides? Time will tell, and evolution will mark the minutes with us.

Chapter Seven

Biotechnology and the Chemical Plow

In 1980, a complex-sounding but nonthreatening research paper appeared with the title "The Herbicide Glyphosate Is a Potent Inhibitor of 5-Enopyruvyl-Shikimic Acid-3-Phosphate Synthase." The authors had discovered the simple cellular mechanism by which a common chemical herbicide killed the plants it touched, and they had also identified the target protein most affected by it. Twenty years later, this kind of work has laid the groundwork for an unforeseen biotechnology revolution and sparked a firestorm of evolutionary change—on an industrial scale. The consequences of this discovery, and the uses found for it by enterprising scientists and entrepreneurs, changed the nature of farming around the world and helped ignite a global fascination with agricultural biotechnology. They also brought 2500 protesters to the streets of Boston in the spring of 2000 and could help bankrupt one of the world's biggest biotech companies. Although the glyphosate story is at its heart an evolutionary one—driven by commonplace accidental selection, and by the brute-force tactics of the new gene tinkerers who rule biotechnology at the dawn of the twenty-first century—it started on the farms of the industrialized world with a sort of agricultural version of tough love: to grow plants you have to kill them.

The Field of Evolution

While using plant poisons on crops seems contradictory, herbicide use has grown ever since the idea of killing plants with chemicals was developed in the early 1900s. The EPA divides pest control chemicals into four broad categories: insecticides, fungicides, a miscellaneous group of other "conventional" pesticides, and herbicides. Herbicides dominate, with a half billion pounds of these chemicals sprayed on crops and roadsides to control weed growth in the United States every year.

Modern herbicides target aspects of cellular metabolism that plants utilize but animals do not; as a result, many popular herbicides are much less toxic to animals than are popular insecticides.

Keeping crop-producing plants free from nutrient-robbing neighbors has always been hard physical work, but herbicides allow a change from classic methods like plowing to those that rely on chemicals to kill the weeds. Although it seems odd that herbicides play such an important role in farming, killing weeds by plowing them under and killing them by contact with deadly chemicals have the same goal: to selectively destroy weeds so that a few favorite crop plant varieties can thrive.

Like insecticides, herbicides can poison more ways than a cheap seafood buffet. Some halt photosynthesis, short-circuiting the capture of sunlight and preventing the forging of carbon dioxide molecules into simple sugars. Some are quite specific in the weeds they kill, whereas more generalized varieties possess the potential to kill even crop plants if used heavily enough.

TABLE 7.1. ELAPSED TIME BETWEEN INTRODUCTION OF KEY
HERBICIDES AND EVOLUTION OF RESISTANCE BY WEEDS.

Herbicide	Year Introduced	Year of Resistance
2,4-D	1945	1954
Dalapon	1953	1962
Atrazine	1958	1968
Picloram	1963	1988
Trifluralin	1963	1988
Triallate	1964	1987
Diclofop	1980	1987
ALS inhibitors	1982	1987

Evolution stalks the herbicide industry just like it harasses insecticides, and more than two hundred different species of weeds currently resist one or more herbicides. Worries over such unstoppable weeds have cropped up ever since the idea of chemical control was widely adopted in the 1940s. Although not quite so speedy as evolution of pesticide resistance in insects, evolution tends to overcome a new herbicide soon after its introduction. The first major chemical herbicide, the chemical called 2,4-D introduced in 1945, induced resistance in weeds by 1954. It took only a decade for a weed to evolve resistance to the hugely popular herbicide atrazine. Picloram and trifluralin had a twenty-five-year run, like midnight screenings of *The Rocky Horror Picture Show*, but were eventually overthrown by different weeds in various parts of North America.

This evolutionary cycle of herbicide deployment followed by weed adaptation complicates a farmer's life enormously. Whether a particular pesticide chemical will do its job represents a critical economic gamble, so that in effect the farmer is betting her yearly salary on whether weeds have evolved yet. Evolution of herbicide resistant weeds represents such a serious problem that many local agricultural boards provide copious guidance to farmers about how to slow it down. The rules they suggest could never be enforced—they are not laws or regulations. But farmers follow them because they make sense and they work, constituting a practical skin drawn tightly over a skeletal framework of evolutionary bones, one whose ligaments tie together evolutionary fact and theory into a well-articulated whole.

The most basic simple rule—don't use the same herbicide on the same fields many years in a row—is as common to herbicide management as is singing in church. Use of the same herbicide year after year would allow a tiny population of resistant plants to grow larger and larger in a particular field. As long as the spraying continues from year to year with the same herbicide, slow-growing resistant weeds escape competition with their nonresistant relatives and multiply. In such a case, resistant weeds are actually being protected by the application of herbicides—protected from competitive elimination by more vigorous weed cousins sensitive to the poisons.

A second rule—to use herbicides with different toxic mechanisms in sequential years—is based on both common sense and evolutionary theory. For example, triazine herbicides like atrazine and phenylurea herbicides like linuron both attack the photosynthetic system, causing the production of highly toxic oxygen radicals inside cells rather than normal oxygen. Weeds resistant to atrazine commonly overproduce an enzyme called super oxide dismutase, which can detoxify the oxygen radicals and turn them into hydrogen peroxide. But super oxide dismutase serves as an effective defense against linuron as well, and plants resistant to one of these herbicides thumb their nose at the other too. So, alternating annually between atrazine and linuron would be equivalent to using atrazine each time. (This type of cross-resistance crops up commonly, not only in plants resistant to herbicides, but in bacteria resistant to antibiotics, HIV resistant to protease inhibitors, and insects resistant to insecticides.)

But if herbicides with different actions are alternated, then the plants favored in the first year are not favored in the second. Selective pressures shift strongly, first reducing the weed population by a huge amount in year one, then switching tactics the next year to eliminate any seeds produced by resistant plants. Only if a plant can resist both herbicides would it and its offspring survive the yo-yo of selection. Such multiple resistance occurs only if single mutations confer resistance to two chemicals with completely independent mechanisms—about as likely as the chance that the special tool you buy to adjust your bike brakes will also work on the bathroom plumbing. Instead, most herbicide-resistant mutations rescue weeds from only a single herbicide type.

It doesn't always work, of course, and resistance—even multiple resistance—has evolved despite this sage and simple advice. The reason is that even if the suggestions are followed, each agricultural field is not a universe by itself; instead, it exists in a patchwork of fields all being treated by independent-minded farmers who may well use different herbicides. Resistant plants selected in your neighbor's field may sow seeds that disperse to yours, and if they used an herbicide last year that you use this year, then resistant seeds may sprout despite all your efforts.

Kansas Evolving

Like an outbreak of swine flu, attempts to legislate the biological process of evolution out of children's minds pick up steam somewhere in the United States every year or so. Recently, in August 1999, the Kansas State Board of Education voted to eliminate questions about evolution from high school graduation tests, thereby dropping evolution off the academic radar screen of teachers and students alike.

Ironically, Kansas residents don't have to go far to encounter the bald face of evolution in action. Falling out the back door would do it for many rural residents. In fact many Kansas citizens, worried about the economic impacts of evolution, are working hard to do something about it. At least ten different weed species in Kansas have evolved resistance to at least one herbicide. A common herbicide, atrazine, has been rendered impotent by four different weeds, including the troublesome redroot pigweed. Another resistance weed, the Palmer amaranth, has been reported in Kansas at a density of over a hundred seedlings per square foot after herbicide treatments failed.

Strategies for dealing with herbicide resistance hold true in Kansas as elsewhere: chemical rotations and switching among herbicide classes top the list. The difference in Kansas? The Board of Education doesn't feel future farmers need to know *why* these steps are important to agricultural prosperity. Since farmers are practical people, they'll use anti-evolutionary field practices when they see them working, but being full of anti-evolutionary rhetoric from education officials can't make it easier to adopt economically beneficial techniques.

The Value of Tolerance

In William Goldman's adult fairy tale *The Princess Bride*, the hero, Wesley (aka the Dread Pirate Roberts), defeats a swordsman, a giant, and Vizzini, the criminal mastermind, to rescue the Princess.

Defeating Vizzini occasioned a tense shell game where two cups, one of whose contents were poisoned, were switched over and over until it seemed impossible to guess the deadly one. Both contenders drank simultaneously—Wesley grimly smiling as Vizzini collapsed midboast.

The Dread Pirate hadn't been worried. Success was guaranteed, for in fact he'd poisoned both cups, having slowly conditioned himself to the poison's effects, rendering his body immune from the poison and from the chance of losing the shell game. He went on to other daring exploits and the Princess's love, still carrying his tolerance of poison but not of boasts.

Like Wesley, modern farmers long ago realized that if they were going to regularly poison their fields, it would be handy if their crops were immune. Because many herbicides inflict some damage on crops, a key role of evolution in agriculture has been the artificial selection of crop plants that can tolerate larger doses of herbicides.

Cicer milk-vetch, a low-growing relative of the pea family, forms a popular perennial planting for livestock forage. However, milk-vetch is sickly while sprouting and easily overwhelmed by more vigorous weeds. Applications of 2,4-D easily control invading weeds, but also kill the struggling, thin sprouts of the milk-vetch crop. But artificial selection on milk-vetch has been successful at increasing the tolerance of this plant to 2,4-D, allowing farmers to grow it more quickly and without expensive, disruptive hand-tilling.

Such programs of artificial selection have been accelerated because use of herbicides has increased so rapidly over the past decades. And in fact, plant breeders have grabbed onto several unique plant attributes to speed up the power of artificial selection, taking advantage of a peculiar part of the normal plant life cycle to increase the opportunities for artificial selection within a generation.

The Last Pollen Alive Wins

New baby plants are born when pollen fertilizes an ovule inside a flower head. Unlike animal sperm, which generally come equipped with nothing but a nucleus, a tail, and a penchant for swimming, plant pollen function as small simple organisms in their own right. They carry only half the DNA content of a normal plant cell, but even so, they retain the ability to sprout and grow once they land on the pedestal-looking stigma of a flower. Having sprouted, an elongating pollen tube tunnels down the length of the stigma, silently racing other tubes of hopeful pollen toward the ovule prizes at the bottom. The first tube to reach an ovule injects that pollen's DNA, winning the race and fertilizing the seed.

Selection can act on pollen, just like on other growing organisms. Stigma themselves actually select among pollen—preventing pollen from the same flower from winning the tube race—and agriculturists have learned to get into the act by dosing growing pollen tubes with chemicals.

Treating a flower with herbicide seems rather like treating a preacher with mace, but mild doses of the herbicide atrazine have been used to select among pollen that carry genes for atrazine tolerance. In this case, the genes that a pollen tube carries are not just the silent legacy of future plant generations being transported down toward the stigma. They work for a living too—by producing the proteins and other structures of the elongating pollen tube. Good genes make for a fast tube, which makes for a successful pollen, whereas pollen tubes poisoned by a dose of atrazine lose the race. Those tubes that can grow the fastest despite the herbicide win carrying with them the genes for enhanced atrazine tolerance.

Once the winning pollen has injected its DNA into the ovule, the new seed that forms has a copy of the gene that confers atrazine resistance, and the plant that eventually sprouts from the seed has the fortune of atrazine tolerance in its every leaf and stem. This doesn't work in animals—you can't foster a musical genius by making sperm swim through a Mozart concerto—but for plants it adds a powerful tool to the bean-stalk magic practiced by modern horticulturists and plant breeders.

This tool has resulted in strong herbicide tolerance in crops like corn. A second tool, born of the need for rapid selection of herbicide tolerance, represents the triumph of an engineering perspective on agricultural innovation—the triumph of brute force evolution.

Biotechnology and Brute Force Evolution

Artificial selection, an ancient tool of both plant and animal breeders, sparked the rise of agriculture as our ancestors of 10,000 years ago chose to plant wheat mutants whose kernels were more amenable to harvest. More recently, John Doebley and colleagues at the University of Wisconsin have traced the evolution of single-stalked corn from a multibranched ancestor called *teosinte*. In Mexico over thousands of years, artificial selection of *teosinte* variants led slowly to the corn we know today. These ancient Mayan horticulturists chose plants with thicker central branches that had larger cobs grown on a single stout stem.

Such selection operates still in modern agriculture, producing virtually all the plant and animal breeds so familiar to pantry and pet store. I am convinced, for example, that asparagus growers have been altering their product so that it no longer produces that peculiar smell after being eaten and filtered by kidneys. This would be a very positive development in the asparagus field, and my direct evidence for this comes from a lack of this smell after recent asparagus meals. It's a delicate subject, one that I've hesitated to broach, but since teenage dinner conversation is hardly any more demure, I brought it up one recent evening. While my children made terrible faces, my wife said very simply, "The alternative is that your sense of smell is going."

I hadn't thought of this. "Well, can *you* smell it?"

Mary shrugged. "About as strong as ever. I can't tell any difference."

"I think maybe farmers have selected against smelly asparagus," I offered.

"Or maybe your sense of smell is going," was the insistent reply.

"There's an experiment we can do, if you're still producing that odor."

Now Mary was making terrible faces too. "This is for the book, isn't it? I'm not having an experiment like that in the book."

So, she refused to help in the experiment, and I still don't know whether artificial selection for asparagus smell has happened or not.

Birth of Brute Force

In the United States, millions of acres of crops, largely soybeans and corn, are treated each year with the herbicide glyphosate. First marketed in 1972 by Monsanto, glyphosate is an organophosphate like many insecticides; however, it sidesteps the nervous system and instead attacks a metabolic pathway not found in animals. Because it has such low toxicity to mammals, birds, and fish, and low environmental persistence, glyphosate is a popular way to kill weeds in fields and vegetation along roadsides (annual sales are nearly $1.2 billion). Weed resistance was rare throughout the 1980s, not described naturally until it appeared in an Australian rye weed in 1996.

It would be just another weapon against weeds, however, if it weren't for the discovery of the mechanism of glyphosate action in 1980. Within a few years of uncovering glyphosate's modus operandi, researchers around the world had engineered plant cells to produce an overabundance of the protein that was the target of glyphosate; the scientists then showed that the engineered plants could grow despite large doses of the herbicide. By putting extra gene copies into plant cells, agricultural engineers mimicked the resistance mechanism mosquitoes had invented against insecticides, and generated crop plants that had resistant metabolisms. By 1987, engineered mustard plants resistant to glyphosate were growing happily, and the seeds of a biotechnology revolution had been planted.

MY SON, TONY, and I have an ongoing debate about computer games, mainly about why they have to be so violent. I keep waiting for a game in which the hero confronts some drooling alien or perhaps a cartoon terrorist with spiked hair using a serious amount of clever negotiation. Or maybe an entire evil league or interstellar empire needs to be overwhelmingly outsmarted. Or maybe some really loud and fast motorcycles are driven like a madman's panic in order to stave off an environmental crisis or prevent a nuclear accident near an orphanage.

Obviously, these are fantasy games.

Fantasy, because they'll never be marketed, and if they are, we won't buy them, and if we do, Tony won't play them. Instead, the games on Tony's wish list employ a very narrow set of human interaction skills to solve problems. Most of these involve triggers of some kind, although some are sophisticated enough to require setting a dial for destructive power before you pull the trigger. We have a family moratorium on the truly and rampagingly violent games, but even the low-key ones run on brute force and virtual adrenaline.

Brute force seems common in many aspects of human life and, as in computer games, we often seem to choose it over more subtle manipulations. This is now true of artificial evolution as well, and it has changed the face of plant breeding.

Genes that confer resistance to glyphosate have been inserted into a farmer's market of crops, instantly creating food plants that can resist the damage caused by chemical herbicides. Because crops will tolerate it, genetic engineering for herbicide tolerance will pump up the use of herbicides—particularly glyphosate, marketed under the name Roundup by Monsanto. It may be as surprising as plaid pants at a golf match that Monsanto also markets the Roundup resistant crop plants all over the world.

Environmentally, Monsanto claims that more use of glyphosate will reduce use of other more damaging herbicides. However, Roundup, already the eleventh most popular herbicide, is the third most common cause in California of health complaints among farmworkers; this is largely because the fluid that dissolves it is more toxic to animals than the herbicide itself. It is also known to drift thousands of feet from the fields it is applied to, damaging native vegetation and protective plantings like hedgerows wherever it lands.

Despite the uncertain logic of environmental benefits of

Roundup crops, 32 million acres of genetically modified crops, mostly corn and soybeans, were grown in 1997. In 1998, herbicide-resistant soybeans alone accounted for over 25 million acres in the United States—a third of the area devoted to that crop. Around the world, more than fifty types of genetically modified crops have been approved for unlimited release, and sale of their seed totaled $1.5 billion in 1998.

Such crops stand at the center of an intense debate between consumers and industrial developers, and between governments touting the benefits of high-tech agribusiness and citizens unsure about the safety of genetic manipulation. The biggest question centers on whether gene-altered crops are any different than artificially selected ones. The USDA, for example, considers genetic engineering just another form of selective breeding, likening the insertion of a bacterial gene into a sugar beet to artificially selecting beets for the same characteristics. Consumer advocates, European governments, and scientific watchdogs like the Union of Concerned Scientists are not yet convinced. They call for more comprehensive testing of genetically modified crops and more careful consideration of potential risks before widespread release.

Some caution seems warranted. Already, genes for herbicide resistance have escaped from their host plants to enter weed species through hybridization with pollen from crop plants. Such genes have also been up your nose if you've walked past a pollen-filled field of modified crops. Each bee carries those genes back to the hive in pollen sacks: they will be in honey. Each seed that escapes the thresher carries the gene into exile in next year's hedgerows or roadside strip of weeds. This uncontrolled movement of engineered genes can have devastating consequences if it gives weeds access to unprecedented levels of glyphosate-resistance genes.

The Revenge of Jointed Goatgrass

Jointed goatgrass is no joke, causing losses of some $145 million annually to wheat farmers. It is such a good competitor that only twenty-five seeds per square yard—fewer than the popcorn kernels under a movie seat—can reduce grain yields by half through choking of wheat seedlings as they grow.

Currently, herbicide treatments help control jointed goatgrass somewhat. By 2002 herbicide-tolerant wheat seed should hit the commercial market, allowing even better weed control since farmers will be able to apply larger herbicide doses. Experimental plots have shown that high herbicide treatments, which are tolerated by the genetically modified wheat, will effectively control jointed goatgrass. For now.

Unlike the situation for soybeans and corn (where few weeds are closely related to the crop plants), wheat has many relatives among North American weeds—including jointed goatgrass. Wheat and jointed goatgrass cross to form hybrids, bastard seedlings with joint custody of genes from both parents. Once these hybrids have grown, they can cross with full-strength jointed goatgrass plants, producing generations of offspring that have mostly jointed goatgrass traits. But selection imposed by herbicide treatments will favor seedlings that carry the wheat-derived herbicide-tolerance gene. The result seems inevitable—the movement of an engineered gene from wheat to weed within a few years. A recent exchange in letters in the journal *Science* makes the same point about the dangers of growing genetically engineered corn in Mexico, where many populations of nondomesticated plants related to corn still survive.

Clearly, this genetic great escape means that a noxious weed may come to cause more harm. Other herbicides besides Roundup might be called for to keep the weed at bay, and maybe the plow will have to be oiled up for an extra run around the field to turn the soil. Once a weed like jointed goatgrass obtains a herbicide gene from a crop plant, it may even pass the gene on to other weeds, and perhaps an entire brace of weeds as tolerant of herbicides as any engineered crop plant will emerge. These super weeds will not stay in the fields of farmers who chose to use the genetically engineered wheat seeds, but they will spread to neighbors who opted not to invest in crop modification, raising the price of weed control for everyone. Already, farmers pay a high price to combat herbicide-resistant weeds—an extra $10 an acre to fight some weeds tolerant of triazine herbicides. How will farmers— those who buy engineered seed and those who do not—share the extra costs of genetic technology escape? We have yet to face this critical social challenge.

WORRY ABOUT SUCH escapes rings throughout the agricultural world. But do genetic manipulations that create this problem differ from Mayan selection of corn thousands of years ago? Or French vintners crossing different vines to produce cabernet sauvignon grapes? Both artificial selection and genetic manipulation are a kind of evolutionary change—the first an accidental mimicry of the forces of natural selection, the second a purposeful mimicry of the imagined power of special creation. These similarities have prompted the FDA to regulate genetically modified crop plants in the same "generally recognized as safe" category it uses for strains developed by artificial selection. Whether genetic engineering and artificial selection are actually the same has turned into a battle for a very valuable high ground.

———

The Breath of Tibet

The major difference between artificial selection and brute force evolution lies in the complexity of interlocking, multiple evolutionary changes that underlies most of the important differences we see between species. Brute force evolution seeks to alter a solitary gene in a small number of plants and immediately produce a desirable trait, population-wide. However, selection almost never works so single-mindedly.

Usually, several genes contribute to any given trait, including many that control the timing of trait expression in an organism. For example, sexual maturity, derived from the action of many coordinated genes, may be expressed early in the life of some individuals, but late in others, as in male salmon that mature after six months at sea instead of a year and a half. Or expression of a trait may occur only in certain tissues, like the red color in Christmas poinsettias, which appears only in leaves surrounding the nondescript flowers. Virtually all genes are orchestrated by other genes, and without such controls, gene expression would be like the whole orchestra playing every note of a symphony at once rather than letting the music flow out one chord at a time. Because of this, a quick fix to a genetic problem frequently is like using the undersized spare tires in modern cars—it will take you some distance, but probably not the whole way home.

THE TIBETAN YAK has a meaty magnificence coated in rough fur and wrapped in stoic power. The term *beast of burden* perfectly suits these biological machines of Himalayan survival: hooves, hair, and lungs that thrive in the thin air, ferrying immense loads through high mountain passes. Tibetan culture has marched on these animals' backs for millennia. Yak oil and butter are household staples, yak leather becomes clothing, yak horns are ground into mortar.

Over the generations, Tibetan yaks have become so adapted to high altitudes they suffer poor health under 10,000 feet. Their coarse hair, hanging in ragged insulating cascades, combines with other features of their physiology to protect them from the rigors of the Himalayas. They have immense lungs—three times larger than similar-sized cows—to pull oxygen from the miserly air. They have less hemoglobin in their red blood cells, and indeed fewer red blood cells, than their lowland relatives. This thin blood allows for a higher ability to withstand temporary dehydration in the dry air and prevents blood cells from being forced out of ruptured capillaries by the high blood pressure required in high-altitude environments. Even the microstructure of their lungs differs. Yaks have thin-walled arterioles in their lungs, allowing better transport of oxygen into their bloodstream.

Yaks have been altered in many ways by artificial selection over the centuries that Tibetans have depended on them. All these changes, all these ways artificial and natural selection have blended to produce animals that can thrive at high altitude, allow yaks freedom on the Tibetan plateau; but, no single change would suffice. Any search for a "high-altitude gene" that would make other mammals well suited to Tibetan rigors would fail—because no such single gene exits.

This example is not unusual. Evolutionary changes brought about by natural or artificial selection usually craft an entire suite of complementary attributes working together to produce an effective change in lifestyle. Seldom does a single genetic change suffice to create a major ecological difference in an organism. If unicorns existed, they would differ from horses in more than a single "horn" gene. And in the heart of the yak, artificial selection must have acted on many genes at once to produce a domesticated beast where none other can live.

Working a Crowd

Artificial selection simultaneously selects for the genes providing a trait plus all the genes that control it. Such selection acts on the whole organism, not just a single gene. Desirable features and any potentially undesirable complications are balanced by the process of selection, an evolutionary personification of the adage "You have to take the bad with the good." As a result, artificial selection must work on the end product of all an organism's genes as they act together to produce stems, roots, and flowers. The sum total is the most important—the way the pieces fit together determines success.

Artificial selection for fundamentally new traits crawls along because it must work on the balance of so many genes at once. While we can see rapid evolution in a generation or two for traits like guppy color, selecting for wildly different characteristics—like single-stalked corn from a candelabra-shaped *teosinte* plant— might take many generations and involve populations in the hundreds or thousands.

By contrast, genetic manipulation can be fast—a trait never before possessed by a plant can be inserted almost instantly—and might involve the production of only a few genetically modified individuals to act as the Adam and Eve of a new crop variety. In the first case, artificial selection sorts among many genetically different individuals, each with a slightly different version of the desired features. In the second case, genetic manipulation generates a very small number of take-it-or-leave-it test plants.

These three differences between selection and engineering—inclusion of the necessary regulatory genes, selection for whole organism success over several generations, and large population sizes with multiple variants—result in very different evolutionary process and potentially very different outcomes.

Suppose your town wanted a practice area for the Girl Scout rifle team. One option would be to develop a rifle range, complete with professional operators, ear protectors, training classes, and unimpeded targets. Different sites could be compared, evaluated, and tested. Regulations could be developed to fit your community's priorities and altered as needed to fit emerging needs. Another way of proceeding involves dropping the guns on Main Street and hoping that everything sorts itself out.

The brute force of genetic engineering is like dropping guns on Main Street. Engineers drop the trait into the organism, loaded with potential impact, and everyone hopes for the best. Usually, no regulatory genes are added except the few that activate the new genes. The complex and species-specific set of genes that control the timing of gene expression—or the tissues in which expression occurs—are in general unknown.

As a result, engineering a new gene into a plant does not guarantee that the plant will embody features planned by the molecular engineers—or envisioned by the stockholders. Engineering focuses on understanding the single genetic mechanism that controls a desired change, not the regulatory machinery necessary to control it. Such engineering solutions might produce the wanted change, but most of the engineering effort has gone into the gene for the trait itself, not the regulatory genes. Because of this tendency, many engineering approaches focus on the rifles, not the rifle range.

The Evolutionary Importance of Failure

Brute force evolution has produced surprising results in one case: Joy Bergelson of the University of Chicago discovered in 1998 that a genetically modified mustard plant began to fertilize seeds of other plants twenty times more commonly than normal. Normally these plants produce pollen that fertilizes their own seeds, but in one mustard variety that had a herbicide resistance gene, use of pollen from other plants increased enormously. For reasons still not well understood, the gene insertion procedure had multiple effects on the plant's fertilization system, effects that had not been predicted by bioengineers.

Thus, when we use brute force evolution, we must build in a stage for culling and sorting. In the final analysis, inserting genes merely engineers variation that can fuel the evolutionary engine. True evolution still must select among these variants. But the biggest difference between natural evolution and brute force evolution lives here. Natural evolution is profligate with its variants and cavalier about failures of individual experiments. For every successful mutation, many, many fail. Evolution in natural populations is a great tinkerer. Selection sorts through hundreds, even thousands, of combinations of different genes to find ones that work best and longest, in different climates and at different ages. The winning regulatory genes and trait genes function together in the plant to produce more seeds than other plants with other gene combinations. The winners declare themselves victors through higher levels of reproduction and survival.

Working within large populations, selection from generation to generation patiently culls the failures, waiting for a truly successful new combination of genes to arise. All mutant genes have to prove themselves in combination with all the other genes. Many combinations are discarded as failures before any are good enough to survive. Trial and error works in this selection process because there are so many gene combinations to choose from, and because the failures can die in droves. Evolution's dustheap overflows with partial successes.

Brute force evolution produces fewer variants for selection to sort among. You wanted the gene to metabolize glyphosate in your soybeans? Here it is—inserted into a single plant, or maybe a small family. How many different ways of inserting genes are tried? How many different genes are employed? How many different ways to accomplish this crop goal are engineered?

When these variants are rare they also become more valuable— and more difficult to discard. The biggest difference between selection of natural variants and a bioengineered crop lies here—in the willingness to discard products if they do not have the desired result. Natural or artificial selection weeds brutally among the partially successful variants, but human engineers do not. Instead, human engineering has a long and successful history of tinkering failures into successes.

If failures are not discarded, if new gene combinations that have surprising or dangerous effects are not detected and eliminated, then modern genetic technology differs enormously from evolution by selection. Ironically, extensive field testing stands nowadays at the fulcrum of the environmental debate. Gene-technology producers, eager to expedite field trials, usually short-circuit normal evolutionary trial and error of their products by producing genetically homogeneous stocks that cannot evolve.

Environmental groups, especially in Europe, object to the whole business of field trials and have taken to sabotaging experimental crops on farms. Ironically, both the bioengineers and the natural-varieties advocates hold evolution prisoner on the sidelines, giving it no opportunity to sort through genetic combinations that work best for the whole organism.

Poisoner's Dilemma

Humans use poisons to control the rest of the biological world, and they've worked, increasing food production enormously and preparing us to feed the 10 billion people we expect within fifty years. Our ability to deliver toxins has grown along with our batteries of genetic and chemical weapons, now focusing on the delivery of exquisitely designed poisons through a mere handful of easily manipulated genes. Evolution has already played a prominent role in our development as chemical farmers, changing the potency of most herbicides and insecticides by producing resistant populations. Evolution certainly will play a more complex role in future development of genetically modified crops.

The question is not whether to use this technology to help double the world's food supply—the burgeoning ranks of humanity leave us little choice. But how do we *best* use genetic technologies? How will the evolution of crops by brute force and the evolution of weeds and insects by normal selection or escaping genes change the face of agriculture? The success of the unexpected is evolution's favorite trick, and engineering solutions must learn to accommodate astonishment.

Chapter Eight

Evolution All at Sea

The first fish you ever caught might still occupy your memory. Yanked erratically and rapidly to shore by small hands, its puny size amplified by excitement and unexpected pride, the fish may have started you on a long career as an angler. Or the biggest one you ever caught might fill this mental niche, having been snared on an otherwise forgotten dawn from a small boat drifting on a glassy lake fringed with metallic fog and cattails. You remember the first tug, deceptively gentle, like the pull of a birthday balloon, and the next, a violent no-kidding snap of the rod that tested your hook and the determination of your thin tackle—this is a story you can tell with verve.

My biggest fish came from Alaska, a door-sized halibut I could barely lift toward the camera. The day sparkled as the ebbing tide and hungry fish snatched my very last cast of the trip. A half hour of careful reeling brought the fish in, but even while my halibut bashed about in the bottom of the boat, my compatriots started in about the *really* big fish caught other places—or other times.

There are three deep-seated reasons that fish caught long ago were truly bigger than the ones we catch today. First, anglers have wonder-fully flexible memories, remembering a fish's size in direct proportion to their emotional attachment to it. Because emotions vary like storm clouds, so too can the memory of a fish—a flexibility that sport fishermen use like a Masonic handshake to identify their own kind.

The brutal success of fishing as an industrial mining activity provides a second, more serious explanation. The fish of yester-year were generally bigger because fishing pressure was less, and fish had a better chance of growing large before we caught them. Today, many commercial and sport fisheries suffer under the cracked whip of heavy exploitation, chasing fish so effectively that once a fish reaches a minimum legal size, it succumbs quickly to a hook or a gill net.

ON A COLD day, in a crowded lobster pot pulled from the depths of the Gulf of Maine, a Down East fisherman would probably find a brace of lobsters too small to keep, and perhaps only one big enough to heave into the tin coffin of crushed ice in the boat's hold. The rest of the catch must be tossed back into the sea's frothy embrace. Luckily, lobsters smaller than legal size easily sur-vive the elevator trip in a lobster pot—and may live out their days as clawed bandits, living off bait meant for their larger brothers. Chicken necks to lobster tails—a transformation even King Midas would appreciate.

However, once lobsters grow to legal size, they have less than a 5 percent chance of growing to the next larger-size class. More than 95 percent of the lobsters taken from the Gulf of Maine are no bigger than the minimum length. They have a life expectancy determined by their growth rate, and like high school students during the Vietnam draft, once they've graduated, they're in mor-tal danger. Oddly enough, legislation sets the minimum legal harvest size less than the minimum reproductive size (unlike the draft), so 95 percent of those lobsters caught never reproduce. This is a perplexing fisheries management strategy that requires the same kind of optimism as Russian roulette.

This high fishing mortality exists for many other species as well. Cod from the coast of Newfoundland or Maine were once such fraternities of monstrous lunkers that the Basque fishing fleet sneaked all the way across the Atlantic for centuries to secretly fish until their holds were full and then dominated the salt-fish markets of Europe. Today cod large enough for reproduction are rare, with spawning aggregations so uncommon that they attract the frantic attention of research fleets scrambling to understand their demise. There is no privacy for cod until the species returns to its former numbers—or goes extinct from shyness.

The Evolution of Heavily Fished Species

Overfishing reduces size by reducing life expectancy, but we can also identify a third reason that yesterday's fish were bigger: they grew more quickly. Fishing tends to target the biggest fish in the school, and extracting the large fish from a fishery has been an evolutionary force leading to dramatic changes in the size of some species. Humans excel so much as fishers that slow growth and maturation at small size represent new strategies that successfully thwart the depredations of hungry fishing fleets. Large fish used to outnumber fishing boats, but in most oceans the tides have turned. More and more, only the small will survive.

I HAVE A hard time not feeling sorry for salmon. They leave home when small and runty—crowds of helpless juveniles flowing downstream in a fingerling waterfall that splashes into the sea—and then face a desperate test against the fiercest predators the open ocean has to offer. Eighty percent of them flail their lives away trapped in nets of hopeless struggle. Always striving against great odds to bypass obstacles and prevail against adversity, they swim back to their native watershed, stubbornly pushing their way up the narrowing stream. Once in the roar of the current, only temporary eddies protect them from the inexorable flow of rapid water that seeks to undo their progress. Some few have the stamina to reach the quieter riffles of their spawning zones, where they fight for space, spawn, and die.

Pink salmon follow this typical life cycle and spend their teenage years at sea, growing from fingerling to a weight of several kilograms. This species has much more regular habits than most salmon, returning exactly every two years, like a candidate for Congress looking for votes. The clock of this unusual exactitude runs so precisely that it has produced two types of pink salmon— the odd-year fish that return on odd years and produce young that will themselves return to the stream in the next odd year, and the even-year fish that do the same in even years. These odd- and even-year fish classes are genetically distinct and have different average sizes.

This precise punctuality means that all pink salmon return to streams to spawn at two years of age, but even so, some fish return bigger than others. And the largest pinks suffer the brunt of the most intense fishing pressure, getting trapped in gill nets that let their smaller cousins slip through. Even recreational hook-and-line fishers try to target larger fish, by fishing at the depth or times that favor big fish bites.

If fishing pressure were low, such size-selective mortality would have little evolutionary impact. But fishing pressure is not low—

pink salmon are affected by the intense effort to supply the market for canned salmon, and estimates suggest that 80 percent or more of the returning fish are caught each year. Size-selective fishing means that only smaller fish make it back to streams—and these fish form the parental stock for the next generation.

In natural circumstances, big fish produce more eggs than small ones and natural selection often favors the largest, most productive individuals. But in today's ocean, big fish have a lower life expectancy, and big salmon find their way into small cans more often then they find their way into streams to spawn. So the lucky salmon, the ones small enough to get away, have an intrinsic growth rate slower than usual. They would have been selected *against* in earlier times—pushed out of the gene pool by fast-growing cousins—but in a world of heavy fishing, these fish with weakling genes have an advantage.

The advantage passes along to offspring, through the generation linkage of inheritance, and therefore the next generation also grows more slowly. Over the past forty years, the average size of pink salmon returning to spawn has decreased 30 percent (figure 8.1). Visible in both even- and odd-year classes, these declines don't represent the simple elimination of the oldest fish (remember, all returning fish are the same age). Instead, declining size represents the evolution of salmon with lower growth rates.

Whitefish Surprise

The same exact scenario has played out in a completely different environment—the freshwater lakes of northern North America. In Lesser Slave Lake in central Alberta, commercial fishing has been intense since the turn of the century. Lake trout were fished out by the 1920s, and lake whitefish declined until the spawning stock

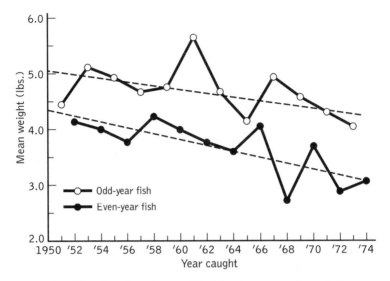

Figure 8.1: The size of two-year-old pink salmon caught at Bella Coola, British Columbia (for odd- and even-year classes) between 1951 and 1974 shows a steady and significant decline. The dashed lines trace the average from year to year.

finally collapsed in 1965. Between 1941 and 1965, the average size for lake whitefish declined, and analysts blamed the removal of older, bigger fish.

But careful fisheries statistics revealed that actually the fish were growing much more slowly than before. In the 1940s a nine-year-old fish weighed about 2 kilograms. By 1960, few nine-year-olds weighed this much, averaging only 1.5 kilograms. Ten years later, the nine-year-old lake whitefish weighed in at only 1 kg—barely the size of the average four-year-old fish from 1945. Young fish still grew just as quickly in 1970 as they did in 1945, suggesting that the conditions for growth within the lake weren't declining. Instead, the fish had evolved a new life strategy—fast growth at first, followed by many years in which weight or length increase only slowly. Slow growth protected whitefish from the nets—two- to four-year-old fish were big enough to be catchable by the 5.5-inch mesh size of the gill nets used in Lesser Slave Lake in the 1950s. But by 1975, only seven- to eight-year-old fish were big enough to be vulnerable—the rest were too small and too thin.

For both pink salmon and lake whitefish, the realities of heavy fishing pressure made a new way of life more successful. The legacy of predation shifted salmon and whitefish evolutionarily, progressing down the fish generations of the twentieth century, whittling away at adult size like Gepetto fixing Pinocchio's nose. We live now in the age of the little fish, when bigger is often a burden.

Two Paths to the Egg

On the tarmac in the cryptic, hazy night of *Casablanca*, Ingrid Bergman balanced between Rick and her husband—between the disgruntled expatriate and the passionate resistance fighter. They were two very different men, one a solitary figure with edges like ripped plywood, the other a silken smooth charismatic leader full of intellectual mission. And when Bergman must choose between them, a rapt audience knows it could go either way.

Famous scenes of dramatic tension pivot on the uncertainties of an unfolding plot, and nothing is more uncertain than which way Bergman will turn. Of course, Rick talks himself out of the running (not the first time silence would have improved reproductive success) but the power of the scene derives from the equal but opposite pull of the two men.

Many animals and plants lay out exactly this kind of mating buffet, where males and females present a variety of tactics for display to potential mates, and where different mating strategies vie for the blessing of natural selection. Which strategies are favored by evolution depends very strongly on the peculiar blend of survival chance and mating likelihood, and when survival chances decline, then natural selection, like the fickle finger of fate, moves on.

After years of heavy salmon fishing, fundamental changes in life strategies have evolved that go deeper into the core of the fish's reproductive strategy, involving cowardice and self-confidence, growth and physiology, fighting and risk taking. It's as if the very concept of salmon manliness has changed.

Salmon return to their natal streams to spawn and die—so goes the plot that everyone knows. But crafted like the polite biography of an infamous politician, this plot hides many fascinating twists and turns, and distorts real intrigue into innocuous channels. True, salmon come back to spawn, but among the males, they also come back to fight. On the way upstream, struggling against currents and the damnation alley of modern hydroelectric plants, male salmon transform startlingly into monstrous-looking fish with crimson sides and humped backs and hooked jaws. Nothing in the journey requires such alteration. Females dress in no such ridiculous costumes. Rather, this gladiator garb cloaks the hope of each male in his contest for females and the best nesting sites.

Once at the quiet, gravel-bed riffles of the stream, male salmon cannot yet relax into a spawning and dying frenzy. Instead, they are thrust into an arena with other males, all vying for the chance to fertilize the most eggs. Some parts of the stream bed are more inviting than others—water flow or gravel condition differs in subtle ways that only salmon can appreciate—and here males take up territorial positions. On station amid the rippling clear waters and the flashing colors of rival males, they wait for females to visit and deposit their eggs. So choosy are females about the males they prefer that in experimental enclosures tended by sharp-eyed fisheries biologists, 90 percent of the eggs are fertilized by just a few males—usually the reddest and biggest males available, occupying the best territories. Like guppies in which sexual selection has dressed males in resplendent colors with bold curiosity in the face of danger, so too does sexual selection favor male salmon that return to streams all large and red, and full of fight.

But two different routes can lead to fertilizations. Fighting for territory and prestige in the eyes of females is the aggressively competitive "Benny and the Jets" way. So-called sneaker males use the second tactic: they hover in the murky shallows outside a good territory, disregarded by other males and ignored by females. When spawning commences, a sneaker male takes advantage of the larger male's distraction to streak into the middle of the happy couple, spraying sperm in the general direction of the female's newly extruded eggs. There the sneaker stays, a desperate fountain fertilizing a few eggs, for as long as tolerated by the rightful male, eventually chased back to the ignominious shallows to wait for another chance. But sometimes his strategy has worked.

When a coho salmon returns to the stream after only six months at sea, fishers and wildlife specialists call him a jack. Smaller than normal, not much more than half the size of a male that waited the regulation eighteen months to go home, he cannot win as a regular hook-nosed fighter, but can easily succeed as a sneaker. Even more extreme, among Atlantic salmon, some males, called parr, do not leave the streams at all. Parr overwinter in freshwater, safe from any marine fishing effort, and stay too puny to excite the ardor of stream hunters.

Small males, of course, produce fewer sperm and fertilize eggs poorly—even a crummy big male can generally obtain more fertilizations. But these sneaker males have a major advantage over their larger brothers—young, and without the burden of many years, they have a higher chance of surviving to sneaker size because they have spent less time running the gauntlet of fishing pressure in the oceans. A male fingerling salmon that never goes to sea would be too small after a year to compete for female attention, but he could adopt a sneaker strategy and still fertilize a few eggs. And his chance of surviving the single year it takes to adopt this strategy far outstrips the survival chances of a male that dutifully goes to sea, struggles upstream, and then fights for territory with his humpbacked brothers. All these challenges make the big-red-male strategy very risky, and one that very few males can successfully complete.

So the two strategies have different payoffs—many fertilized eggs for big successful males and few for sneakers. But they have just the opposite chances of success—low for red males and high for sneakers. In one careful study of the total expected payoff—the number of fertilized eggs multiplied by the chance of success—Mart Gross of the University of Toronto showed that the two strategies had about equal chance of reproduction. The high payoff of territorial males just about balances the high survival of the sneakers. Such equivalence explains why salmon maintain such variable spawning strategies within a species, why natural selection has not eliminated one or the other. A *best* strategy does not exist.

But this equivalence depends on the fate of males that go out to sea, and their fate has lately become much grimmer. Male survival, low under normal circumstances, collapses once males enter the oceans crowded with fishing boats. This shifts the balance between sneaker and big-red-male strategies in favor of little guys, because in today's fishery the safety of staying small pays off even more, and this strong selective force increases the proportion of sneaker males in salmon populations.

————

Sneaker Selection

In a normal coho stream about 25 percent of the males produced turn into jacks. With increased fishing, this number should rise as selection favors males that avoid the part of the regular salmon life cycle that entails the most risk. But other differences between small and large males demand attention before we can predict their evolution. In some cases, only the biggest males have the energy to fight their way upstream—a male salmon expends nearly 50 percent of his body mass to force his way home. Small jack males may not have the stamina for this trip so the strenuous upstream journey selects for larger males. It has proven very difficult to include this feature in a prediction about the evolution of salmon mating strategies. But a way around that problem exists: considering populations in which the stream-fighting part of the life cycle has been eliminated. In those cases, rapid evolution to high proportions of sneaker males becomes the clear evolutionary prediction.

Salmon hatcheries, once the cornerstone of salmon recovery efforts, serve as the stage for this experiment. Usually situated near the mouths of rivers, hatcheries grow fish to the fingerling stage for release. The adult fish, when they return in a few years, go right back to the hatchery, a display of loyalty that has endeared salmon to many fisheries' biologists. But the fish do not have to fight their way upstream, and so the selection for large male size has been eliminated. In these hatcheries, jack salmon have the advantage, and as a result, in some hatcheries, jack salmon have quickly become the dominant male type among returning fish. In some cases they make up over 75 percent of the individuals returning, despite efforts of hatchery managers to slow the trend by choosing big males for breeding. The resulting evolution produces exactly the fish the fishery does not want—small ones with lower muscle mass and too much testes.

Sex Change and the Successful Fisherman

We use the sea as a grocery store so commonly that it may come as a shock to realize that more than 50 percent of fish stocks have been overfished or fished to capacity. Fisheries decline when the death rate has increased due to fishing, disease, pollution, and other factors while the birth rate has decreased because of a smaller number of spawners and poor growth. And under these new environmental conditions, the life cycles used by many current marine populations, so appropriate in the past, fail to successfully navigate seas with a higher death rate and lower birth rate.

Changes in salmon reflect new evolutionary pressures exerted on fish by the world we humans have made. Similar changes run through other fisheries as well, reflecting massive differences in lifestyles required for success in the modern world of stainless steel hooks and factory ships. In at least one case, this has led to a fundamental change in sex determination for the hunted species. Because of overfishing for a once abundant northern shrimp, sex change is now a vanishing way of life.

The cold-water rock shrimp, *Pandalus borealis*, lives in the northern Atlantic, where Swedish and Danish fishery increased tremendously in the late 1950s and 1960s. As in many fisheries, shrimpers target the largest tails, and the proportion of shrimp over 80 millimeters declined from 44 percent in 1950 to 14 percent in 1962. Unfortunately for *Pandalus borealis,* almost all the large shrimp are female, because in this species each individual undergoes a sex change sometime after the first year of life.

Born male, every rock shrimp functions as a male until it is 75 or 80 millimeters (2.95 to 3.14 inches) in length. Then it abruptly changes sex to a mature female. Females hold on to eggs under their abdomen, ventilating the egg mass with oxygenated water using powerful oarlike legs. Only large females can carry large batches of eggs, whereas even the shrimpiest male will serve to fertilize a big clutch. For this reason, it makes sense in this species for small individuals to perform fertilizations, and for them to switch to being female when ready to carry large egg masses.

But if the fishery takes all large animals, then it leaves few females to provide any eggs, and there are few advantages to waiting until after the male years to become female. A better evolutionary strategy—some individuals could reproduce as females as early as possible, before the fishery claims them—has quickly arisen.

Before 1954, there were no females less than 75 millimeters long—all smaller individuals were male, doing their duty in the land of giant mates until they themselves converted to shrimpette. Soon after 1954, coincident with the increase in fishing pressure, small females began to appear. By 1958, 10 percent of rock shrimp under 75 millimeters were females. By 1962 this figure had jumped to 30 percent. Quite likely, those pioneers that converted to females early—or were born female—had greater reproductive success than the normal sex-changers. Greater reproductive success meant having more offspring, and as a result, the fraction of non-sex-changers began to increase in the population. A complex life cycle shifted to a simpler one because of the huge selection against large individuals.

Written deep in the life cycle of these rock shrimp lies a lesson in caution: evolution acts on the whole life cycle, not just a single piece of it. True, in some cases, natural selection acts so strongly on a signature feature—like whether or not a finch has a strong enough beak to survive a drought—that we talk about the evolution of this one trait. But even then, a trait must function in the context of the rest of the organism's morphology and life cycle.

Once born onto the planet, we spend our entire lives in its thin biological wrapping, and human industry affects an organism throughout all stages of its life cycle. The species around us need to live their whole lives under our new management. For them it's a gambler's world—it always has been—and the evolutionary process acts as a blind bookie totaling up the scores of every individual's approach to life. Environmental change affects the underlying rules for successful life cycles, changing the odds and the point spread, the winners and losers, with the result that the evolution of life cycles—or their extinction—becomes inevitable.

Evolution and Plasticity

The story of evolution needs characters and dramatic plot lines, and good tales have a way of sinking into the fabric of evolutionary thinking. But alternative explanations need exploration for even the best stories, and evolutionary biologists have been relentless in pursuing these.

Geoff Trussel faced an uphill battle. As a graduate student staring at a well-known evolutionary story, he thought he could see a very different explanation, but needed hard proof before toppling any legends. The story revolves around change in a snail's shape after the invasion of the predatory green crab on the East Coast. No one doubts that the shape change happened. Pioneering work by Robin Seeley showed that shells collected before the green crab invaded New England in the late 1800s were thin and pointed, just perfect for crabs to peel and consume, but that after the crabs swept into the area, snails became thicker and rounder, and much more difficult to eat. Seeley assumed this derived from evolutionary change—selection for snails that resisted predators—but Trussel had another possibility in mind. He caged snails and crabs together, separated by a mesh barrier like unfriendly dogs in a kennel. In close proximity to their predators, the snails grew a tougher shell. Each individual could change in response to a selective agent, and this did not represent genetic evolution of populations.

In the late 1700s, Jean-Baptiste Lamarck championed the idea that organisms could change in response to their environment, and that they would pass along their changes to their offspring. This Lamarckian view of evolution died out, succumbing to the Darwinian view that selection works on ingrained, genetic traits. But we know that many organisms respond to their environment by altering their physiologies. My niece played lacrosse for Colorado College, a beautiful campus nestled at the foot of Pikes Peak. The campus sports terrific athletic facilities, but with the mile-high altitude comes thin air. Humans respond to thin air by increasing the amount of hemoglobin in our blood, making it easier to gulp oxygen when scarce. This acclimation to environment works so well that my niece's team could run rings around any team that came to Colorado from a sea-level college—a home team advantage that left opponents gasping in the discount air.

The home team advantage in New England snails came from their former home in Europe, where both snails and crabs had lived for a long time. There, the snails had evolved a strategy of growing quickly by making only a thin shell in those places, like rocky promontories, that crabs could not inhabit. Where waves were gentler, crabs were more common and snails grew more slowly, investing more in the thick shells they needed for protection.

These snails were absent from New England until the glaciers of the last Ice Age retreated to their Canadian winter homes, exposing an inviting rocky shore. Somehow, the snails got to New

England early, but the crabs did not. And once here, the snails grew thin and fast and unconcerned about the subtidal local crabs. Only after the European green crab rode to New England on a ship did the snails need their ancestral ability to respond physiologically to the presence of crabs—and then they changed almost instantly.

Not evolution by natural selection, this represents physiological change through acclimation, the ability to respond to environment by changing some aspect of an organism's physiology or lifestyle. Although ultimately produced through natural selection, the ability to acclimate drives substantial ecological change in natural communities. It even extends to some fundamental properties like sex.

Climbing the Social Ladder

Among the crystal waters of the Caribbean, blue-headed wrasses swim in short hops and erratic jumps, studied intently by a shadowy figure up above. Picking tiny animals from the water currents for food, wrasses patrol a particular part of a reef, watching for food, danger, and potential mates, unaware that floating marine biologists spend each summer intently documenting their amorous lives.

Bob Warner has snorkeled in Saint Croix so many years he knows the reefs by heart, and every afternoon he and his students take to the water to watch and record. Along a short stretch of reef, usually just the size of a small car, they focus on one particular fish. As large as a banana and helmeted with a bright, cobalt head, this male controls the territory and the smaller, yellow-striped females in the vicinity. Wrasses schedule their mating for late afternoon almost every day, when the male and all his females strew eggs and sperm into the water with abandon. But Warner, through careful identification of all the local fish, knows that this male once sported a female's stripes. When the previous male disappears through the accident of predation or storm (or Warner), the largest female in the cluster begins a fast change from head female to blue-headed male. She changes behavior within seconds—and mates as a male that very day. It takes five days more to actually produce sperm, and by then her reign as king begins.

This beyond-the-call-of-duty form of physiological acclimation—realizing there is a vacancy in the post of local male, and changing sex to fill it—commonly occurs in sequential hermaphrodites like wrasses, groupers, and some crustaceans. Could some of the evolutionary change in rock shrimp stem from a hidden ability to gauge the number of local females and change sex when fishing wiped them out? A few recent studies of the environmental control of sex in shrimp hint at such possibilities. Could male salmon take up the jack or parr strategy more often these days because they are responding physiologically to some recent change in stream environments? Also possible, since we know that the fastest-growing salmon fingerlings most commonly become jacks or parr, and that low salmon spawning densities may reduce competition for food among hatchling fish. Some think that this is one reason jack salmon are so frequent at the return gates of salmon hatcheries—fat hatchery fingerlings turn into jack males more often than if they had to make their own living in natural streams.

In fact, marine environments seem replete with examples of rapid acclimation of individuals to changing conditions, changes that might usurp some of the need for genetically based evolutionary change seen in other habitats. This does not obviate the role that evolution plays in recent ecological changes in the sea—but it may mean that the key evolutionary events occurred long ago, when selection fine-tuned the ability to judge and respond to environmental change. Unlike the Lamarckian model, these physiological changes do not automatically pass to offspring. Instead, the *ability* to change sits intrinsically within the genetic makeup of a species and in Mendelian fashion is passed on to the next generation.

Such responsiveness to environment emerges most clearly in behavioral traits like feeding method and movement. However, in simple animals cultural evolution does not occur, since there is no way for a baby shrimp or fish to learn from its parents, and information does not flow down the generations. In more behaviorally advanced organisms, when offspring can learn from mom or dad, then cultural evolution can play a strong role. The premier marine example is the whale.

Pushing Whales Around

In Hawaiian legend, *Ka-ehu* was a small, yellow shark god, raised by his parents on the south shores of Puna on the island of Hawaii, and a well-known resident of the rippled sandbars of Pearl Harbor. His world, crowded with other powerful sharks—like *Ka-moho-alii*, brother of the volcano goddess *Pele*—held a strong connection between the powerful sharks and the people crowding the coral shores. There were less savory characters too, like *Pehu*, the man-eating shark of Maui, and on the island of Molokai the shark god *Kauhuhu*, whose ripping revenge created the legend of *Aikanaka*—the place of the man-eaters.

Indeed, Hawaiian legends are replete with the names and deeds of sharks. Man-eaters and gentle helpers, gray rockets and shy yellow reef dwellers. And not just sharks swam there. Populating the cultural diversity of Hawaiian legend and song are many other members of the marine cornucopia, animals and plants whose names flow as liquid as the sea. *Lau-wili-wili-nuku-nuku-oi-oi*, the long-nosed butterfly fish (literally, the long-beaked fish that floats like a leaf), cruised the legends in dedicated pairs like it swam the reefs. Porpoises and manta rays leap into legend as *nai'a* and *pe'a*, and at least thirty-three other species of fish populate the second chant of the *Kumulipo*, the Hawaiian hymn of creation.

But missing in this hymn to the Hawaiian past is one of the most conspicuous and majestic of all marine animals, the humpback whale. This absence seems odd given the current role of humpbacks in local Hawaiian affairs: the long-finned families crowd Hawaii's shallow turquoise seas. These mammalian steamships are so common that a booming industry has developed to take tourists to watch them. So frequently do they throw themselves into the air that the thunder of their waves can be heard from shore. I've been on beaches where the sight of leaping whales is so typical that boisterous children are more impressed by an indigo jellyfish washed up on the opaline sand than they are by the whales tasting the clouds offshore. Where were the whales in old Hawaii, and how did they arrive? No petroglyph has chiseled their history into the Hawaiian basalt, no one carved their awe into wood or stone. No certain solution to this enigma has ever been proposed.

"Uncle" Charlie Maxwell, a minister of old ways and new, sat in a white wicker chair on a moonlit night listening to the waves and a luau compete for musical melody and told me what he thought. He insisted that humpbacks were the *kohola* of the old chants—that they were there all along and played a key cultural role. There were places where you would go, he said, to watch the whales move through the islands every winter like they do now. They were here and they were a part of us.

But Lou Herman of the University of Hawaii suggests a different possibility—that the humpback whales discovered the Hawaiian archipelago only recently and that they were not in old

Hawaii at all. The whales of legend, *palaoa*, were the rare sperm whales whose scimitar teeth ornamented the leis of the powerful, *lei niho palaoa*, or were the heart of graceful curved ivory fish-hooks, *maku palaoa*.

Herman posits that whaling along the west coast of North America drove humpbacks to seek a quieter shoal. If so, humpbacks discovered Hawaii at a bad time—the height of the whaling days, when ships crowded Lahaina harbor like sailors would a tavern with free beer, and whaling crews were bought, bribed, or stolen for the horrible trip to the Pacific grounds of the sperm whales. The newspapers recorded the presence of occasional whales, and the bloody result. Small boats chased mother-calf pairs, wounding the calf to keep the mother from fleeing, and selling the kill to an idle whale ship for half the oil. From the 1840s through the late 1860s, a small shore-based whaling industry grew up among the diffident whaling ships from distant seas, though not many whales were taken. At about this time, the price of whale oil crashed on world markets, victim of this new thing called petroleum. The doom of the whaling industry signed a permanent lease in Hawaii for the humpback whales.

This still does not explain where the humpback whales came from, and what brought them like eager tourists to Hawaii. Previous to their extended Hawaiian holiday, they must have had homes elsewhere.

Humpback whales wander. They spend their winters in shallow warm waters, breeding, fighting, and judging the song contests of the local males. But they spend summers, a serious time of overeating, in the gray larders of polar seas, which are thick with the small fish and crustaceans that constitute the humpback menu. In between, they swim with slow strokes of monster tails back and forth, crossing the planet as easily as a parking lot. These movements, neither random nor capricious, represent migration routes followed faithfully year after year.

Whale DNA confirms this part of the whale lifestyle. California whales differ slightly from those that winter in Hawaii and summer off Alaska. They also diverge from their cousins in the southern Pacific or in other oceans. Whales can swim around the world on a whim, but they stay faithful to particular breeding reefs and return to familiar feeding grounds year after year. At rare intervals, they claim a new migration route—otherwise, only one migratory group of humpbacks would exist in the world, not a dozen. An old clan, humpbacks have graced the Earth three times longer than humans, and they change their routes as seldom as a river.

Yet, these whales did find Hawaii at some point, and their DNA holds a clue. Genetically, humpbacks in Hawaii are the strangest of any in the world. In most of the world's humpback populations, a good deal of genetic variation exists among different families. In fact, mitochondrial DNA—a genetic marker often used to trace family lines from mother to daughter—retains about ten times as much variation in humpback whales as in humans. But in Hawaii, the rules are different. We've uncovered little variation in mitochondrial DNA in Hawaiian whales, and family groups there are so closely related we can distinguish few whales from one another without more sophisticated genetic probing.

This evidence shows that the Hawaiian humpbacks descend from just a few ancestors who stumbled upon the archipelago, and that only a very few whale families successfully colonized the Hawaiian Islands. When did this happen? While the evidence is far from conclusive, the most likely interpretation of the genetic data indicates that humpback whales migrated to Hawaii a hundred years ago, when the Lahaina fleets first noticed them.

Genetics also shows us where the ancestral Hawaiian whales

come from—not the southern Pacific (starting place for the Hawaiian people themselves), but from the west coast of North America. North American humpbacks ordinarily move from near the Farallon Islands off San Francisco, south to the winter warmness of Baja California. Some whales shoot out to the isolated Islas Revillagigedo to feed, others may travel far north of San Francisco, but the main population today moves north and south in a predictable annual migration that brings them to Baja for the winter. Although genetically unique, these whales represent the closest genetic relatives of Hawaii's population.

Here is the puzzle: If these whales have a regular route that persists today, what displaced the Hawaiian ancestors, sending them questing out to sea for a new winter habitat? That something may have been hunting. Along the west coast of North America in the mid-nineteenth century, the bloody carnage of active whaling may have chased migrating humpbacks far offshore. Whale migration patterns, affected by local conditions, respond to simple environmental changes, like the loud noises that cause gray whales off the California coast to veer suddenly farther offshore. Continual whaling along the humpback migratory route may have forced them into the middle of the Pacific, where a short swim could take them into the arms of the Hawaiian archipelago. Once hunting ceased, through protective measures imposed by the International Whaling Commission in 1966, the Hawaii whale population began to grow, and became the tourist attraction and marvel of natural power seen today.

Uncertainty surrounds this tale. Were humpbacks truly absent from old Hawaii? Did human whaling activities truly cause the shift in migration route? The genetic evidence supports the recent arrival of the Hawaiian humpbacks, suggesting that human-induced change has been a recent driver of whale migration pattern, leading to a whole new population of whales thriving in the paradise waters of the central Pacific.

Evolutionarily, the whales haven't changed—except in one important way. They now possess a cultural imperative different than their forebears. The migratory pathway passes down from mother to offspring, like hearth stories or how to avoid bullies at school. One generation inherits this information from another almost as precisely as if it were actually encoded in the whale's genes. Thus, the shift in whale migration represents a cultural mutation, an increase in the behavioral variation of the species and an increase in the fuel of evolution. Selection may have stemmed from the lack of avid hunting in Hawaii compared with the west coast of North America. And cultural inheritance may have been in the training a calf receives as it follows its mother for several cycles of the migration pendulum. The ingredients are here for evolution to act: variation, selection, and inheritance. In this case, DNA does not encode the trait—instead, behavioral tradition is the clay tablet on which the trait is scribed. Nevertheless, culture can evolve when all three evolutionary elements exist. On land, we easily study cultural evolution in humans, chimpanzees, birds, and other social animals. Whales may be the crowning example in the sea.

Chapter Nine

Are Humans Still Evolving?

W e like to pretend that universal rules don't apply to us. Richard III proclaims "I am myself alone," and rampages his kingly way through two Shakespearean plays broadcasting untouchable contempt for human civility. Closer to home, rule-skirting, king-pretenders carry fifteen items into the twelve-item express lane at my local supermarket—maybe rationalizing that nobody around them can count. I hold my breath, hoping against hope the manager will come over and haul them off to express-lane purgatory. But it never happens—the express-lane limit is not for them.

People have generally preferred exemption from evolutionary rules too. The nineteenth-century storm over evolution crashed onto English society at the first suggestion that evolution's process fully affects us humans. Darwin sought shelter from this weather under an umbrella of tact; in *The Origin of Species*, he barely mentions humans at all, except as perpetrators of artificial selection. But no one was fooled by such blatant omissions, and whether or not humans were cannon fodder for the artillery of evolution drove the social debate then, as it still does today.

The initial argument was philosophical and theological. Surely, evolution exempts us humans by virtue of our unique perch at the top of life's hierarchy. This Olympian viewpoint, rejected early on by evolutionists, quickly gave birth to another notion: perhaps, after all, we are exempt for solid biological reasons. Because our intelligence clearly allows us to overpower nature, perhaps it allows us to overpower natural selection as well. Our hegemony over nature being cited as evidence, by reason of sanity, we were found not guilty of evolution.

IT'S A PARENT'S job to prevent evolution, especially on cold New England mornings, often by finding their children's misplaced gloves. Gloves display a mysterious tendency to vanish on the coldest days, and our parental responsibility of buffering natural selection, letting its sting soak through to our children only in pragmatically educational ways, requires we block its path by finding the gloves on mornings when little fingers might freeze.

Nor does our evolutionary buffering stop at the nursery door. I sit at the keyboard, staring through contact lenses and wearing a long-sleeved shirt to keep out the New England chill. I've killed very little game this week, and been generally unprepared for wild predators stalking the icy lawns. And except for vaguely noticing the stock market oscillate, I haven't been hoarding much for winter hibernation. In short, as a traditional mammal facing an impending winter, I'm a disaster.

So far, though, it hasn't seemed to affect my reproductive success—in fact, the contacts may have helped. But has our brazen defense against natural selection brought human evolution to a halt? Am I polluting the human gene pool with weak vision genes?

Francis Galton, Darwin's cousin, and one of the originators of the mathematics of statistics, worried enough about the lack of selection for intelligence in the nineteenth century that he founded the Eugenics Society, an implicitly racist social club dedicated to pairing up highly intelligent people. Although finding Galton's solution outrageous, many famous twentieth-century evolutionists agreed that human evolution had dramatically slowed and decried the predicted accumulation of harmful mutations in humans protected from selection by technology. Julian Huxley asserts in the 1953 book *Evolution in Action* that "the more elaborate social life is, the more it tends to shield individuals from the action of natural selection," although the quote makes it clear he never lived in a fraternity house.

When we provide medical treatment to the injured, give food to the hungry, replace brute muscle with John Deere tractors—the argument goes—we prevent selection from weeding out a myriad of weak or physically imperfect individuals. Because we can clothe ourselves in winter, feed ourselves from storehouses during droughts, predict tsunamis and the paths of hurricanes—the argument continues—we break the link between physical variation among different humans and differences in reproductive success.

As I stare at this computer screen through my artificial lenses, I can imagine that my shortsightedness would have handicapped my success in Paleolithic society. Maybe I'd have ended up chipping stone tools in the back of the cave and squinting in vain for a chance to breed when the sharp-eyed hunting party left. In those times, my blurry-eye genes would not have been passed on, but now my society affords me the chance to see no matter how crummy my worldview, and I pass my bad vision genes—as well as other, better qualities—to my children.

I do other bad things to thwart selection, like buying lactose-reduced milk. This prevents unpleasant stomach cramps and a rush on the bathroom after New England clam chowder. But it also buffers my family and me from selection against our genetically based inability to digest lactose. Mother's milk has a high concentration of a sugar, lactose, that can't be absorbed directly by the human digestive system. Instead an enzyme called lactase breaks the lactose into two simpler sugars easily absorbed by the gut wall. Without the enzyme, lactose accumulates in the gut, leading to stomach cramps and diarrhea. Infants, young and reliant on mother's milk, produce prodigious amounts of lactase, but in most of us the lactase-producing gene shuts down as we get older, no longer producing the enzyme or helping us digest the lactose load in a glass of milk.

But some lucky people—about 50 percent of Americans—can go on drinking milk well into adulthood with no ill effects. In particular, people of northern European extraction possess a variant of the lactase gene that does not shut itself off in older children or adults. Presumably, the high dairy content of the diet in northern Europe selected for a population with a mutant lactase gene, one that allowed people to consume milk into adulthood. Only one other population possesses a similar ability—pastoralists originating in north and central Africa, who herd cattle and mix cows' milk and blood for dinner.

Genetic discoveries about adult lactose intolerance have led to Lactaid milk, which is treated commercially with extracts of bacteria highly concentrated in lactase. This white elixir now makes it possible for people whose lactase genes turned off even earlier than mine to continue drinking milk. In my family, when we buy this milk, we protect ourselves from the natural selection of stomach cramps and pollute the gene pool.

Snipping the link between physical variation and reproductive success stalls the engine of evolution by wrecking the pistons. So, with the engine dead, how can human evolution continue? Maybe it doesn't—maybe technology has stopped human evolution cold. A more complex answer concludes that human physical evolution may have ceased, but our mental evolution—the ideas that drive our culture and technology—has accelerated. Christopher Wills drives this point home when he writes in *Children of Prometheus*, "Selection pressures have actually shifted from the visible to the invisible, from forces that test the resources of our bodies to forces that test the resources of our minds." We will return to the evolution of ideas in the next chapter, trying to understand if evolutionary thinking helps us understand the breakneck pace of human cultural change. But I believe that our technology has in fact failed to halt human physical evolution, for the troubling reason that the human condition is not so universally rosy after all.

A World of Difference

Dr. Richard Bail put his Boston medical practice on hold and flew to Zambia in December 1999, when the AIDS epidemic had gathered itself into a raging tidal wave and the world's attention was finally being directed toward Africa. His mission for the World Health Organization was to help understand the costs and options of different medical approaches to AIDS and to take a hard and practical look at the realities of providing that care. He took leave from a famous Boston HMO, with a generous drug policy that charges people with AIDS $30 a month for $1500 worth of triple-drug therapy to control their HIV infections, and stepped into a country where 1 million people—one in ten—have untreated AIDS, a place the disease has overwhelmed like the hot sun on a parched plain.

In Zambia, antiretroviral drugs are rare. A dose of simple AZT to prevent HIV transmission from a pregnant mother to her infant, which costs $50, is nearly eight times the annual per capita expenditure on health care by the national government. In this economic environment, HIV rages unchecked. Triple-drug therapy for a month would deplete the entire wealth of a prosperous family. Dick Bail found what he had feared: having AIDS in Zambia and having AIDS in Boston were as different as vultures and hummingbirds. The differences in wealth and access to health care stagger the imagination.

Any arguments that we have conquered evolution ignores what was made plain to Dick Bail in Zambia: the human condition varies enormously from place to place. Even in twenty-first-century society, poverty remains endemic. Although wealthy nations provide good health care, poor people in poor nations lack stable nutrition and basic medicine. In the United States, the per capita expenditure on health care in 1998 was $4094; in Zambia, it was $6.54. That same year, infant mortality stood at 109 per 1000 in Zambia but at less than 10 in America. UNESCO reported in 1993 that 3 million children a year die of diarrheal diseases, which are those associated with poor water, bad sanitation, and insufficient nutrition. According to the United Nations, there were 800 million chronically underfed people in the world in 1990, and about the same number living in poverty. Poor access to basic health within the developing world creates a dichotomy in which ailments that I can shrug off with a prescription may strike fatally in a family somewhere else. A Tuscan proverb advises, "The best remedy against malaria is a well-filled pot," an observation that rings all too true in the shantytowns of the world.

Even basic access to clean water varies enormously from society to society. In the 1980s I hitchhiked through towns along the north coast of Jamaica that were all rich because they had a tap connecting them to a purified and treated national water supply. But the 1990s saw Jamaica's national water system collapse, returning towns to dependence on local streams. Standing on the typical concrete bridge over such a lifeline in 1995, I watched water run darkly through eroding gullies, washing down animal waste and precious soil from upland farm plots. After visiting its gift of disease on the towns, runoff was killing the reef too, thus doing double service of bringing illness and destroying the fish supply. Such scenarios abound in the developing world: according to a 1998 World Bank report, less than half of Nigeria's 111 million people had access to clean water, and only about a third had access to sanitation. And as the cost of basic medicine escalates, the economic gap between people with and without adequate food, health care, or simple water widens to a chasm that most can't cross.

When disease organisms evolve, matters worsen. Cheap tuberculosis treatments become astronomically expensive if you come down with a drug-resistant strain. In the 1990s such an epidemic began raging in Russian prisons, an outbreak that could have been quickly halted with the right drugs. But these drugs cost up to $100,000 per cure, a price that places recovery on the other side of the chasm—across the divide that separates economically realistic cures from those available in wealthy nations. Closer to home, drug costs constitute a primary burden for many senior citizens in the United States. Because drug costs rose by almost 15 percent a year in the 1990s, the Commonwealth of Massachusetts, wealthy by any standard, currently struggles to find a way to provide critical drugs to cash-strapped pensioners.

These snapshots caricature only crudely the drama of life across our varied globe. Even so, they exemplify how a disease that plays no role in survival in the United States could change life expectancy somewhere else—how survival may depend on the right combination of genes and environment, and how selection still operates in the human species. While the human condition varies so enormously from place to place, the engine of human evolution will turn.

Useful Variation

Theodosius Dobzhansky, one of the century's leading evolutionary geneticists, emphasized the connection between poor conditions and continued selection in his 1962 book *Mankind Evolving*; at the time, however, he could illustrate his point with few examples. The mechanistic links between genetics, disease, and selection required the intervening decades to come into crisper focus as information about human genetics crystallized. Today's modern toolbox of molecular genetics has opened up much of the human genome to scrutiny, discovering a swarm of variation among the parking lot of human genes. In the three billion or so positions in the DNA in your cells, you most likely differ from your neighbor or your spouse at one million places. Most of these differences will not be in important regions of the genome. They won't even be outwardly visible, but instead be hidden in the vast tracks of junk DNA, like a typo in a newspaper no one will notice. Other changes may have landed squarely in areas of critical function, but may cause no harm due to their chance invisibility to the cellular machinery. Nevertheless, among those million differences, several hundred probably impact human life. A few could even be pivotal to survival.

Among these examples nestle some of the most common and well-known human hereditary diseases. Intense study of these diseases has mined a wealth of information about their impact on health problems, and in rich nations, there is a good chance that gene therapies can reverse the terrible effects of these inborn genetic errors. In many respects, these diseases represent the cruel edge of human genetic variation—the legacy of damaged DNA and consequence of damaged cellular mechanisms. But a persistent mystery remains: Why they are so common? Julian Huxley was just one of many evolutionists to articulate a popular answer—in humans, natural selection has stopped because of our culture, and so mutations are no longer weeded out. "There can be no reasonable doubt," Huxley concludes in *Evolution in Action*, "that the human species is burdened today with many more deleterious mutations than can possibly exist in any species of wild creature."

But modern genetics suggests something different—that at least some of the most common genetic diseases show a quilting of past selection, and continue to be under strong selection now. They do not prove human evolution has halted, but quite the reverse.

The Shadow of Cystic Fibrosis

Sarah Walters, a physician and health consultant in England, remembers a surprise holiday to Scotland when she was twelve years old, made all the more mysterious because her parents took her out of school for it. It was a lovely vacation, in a motor home, to Blaire Atholl castle and Loch Lomond and Loch Ness. Only later did she discover the reason for the trip. When Sarah had been diagnosed with cystic fibrosis at ten, her parents were told she had but four years to live. The sudden vacation was her parents' gift—they thought she should have one good holiday.

Until recently, this family reflex was normal, justifiable because cystic fibrosis was typically and rapidly fatal. The *Almanac of*

Children's Songs and Games, published in Switzerland in 1857, typifies the fear:

> *Woe to that child*
> *when kissed on the forehead*
> *tastes salty.*
> *He is bewitched*
> *and soon must die*

One of every 2500 children born to Caucasian parents "tastes salty,"—a reference to the tell-tale symptom of excessive loss of salt in sweat—and up until about 1950, these children usually died in infancy of cystic fibrosis. Although modern antibiotics and physical therapy currently stave off the disease for decades, with average life spans of about thirty years in 1995, cystic fibrosis still stalks hospital wards in a nursing gown of blocked airways and failing lungs.

A disease of the lungs, gut, and pancreas, cystic fibrosis has a wide variety of serious symptoms including thickened mucus accumulating in the lungs and digestive system. The failure of the lungs to clear themselves of debris sparks a series of infections that recur with alarming frequency. With such strong selection against it— such a high death rate—why do cystic fibrosis genes occur so often, with one in every 50 Caucasians carrying them? The evidence now suggests that recent and perhaps ongoing selection bears heavy responsibility for the high frequency of the cystic fibrosis mutation.

IN NORMAL LUNGS, workaholic cells form a continuous film over the airways and function to transport salt from the blood stream to the lining of the air passages. Because these cells also leak fluid like a dropped carton of milk, water normally oozes from within the cells onto the lung lining to dilute the transported salt. This functions as a well-regulated mechanism of lung cleansing. By loosening the lung mucus, ample fluid flow allows millions of beating, hairlike cilia to transport the mucus out of the lungs, allowing the lungs to clear themselves of dusty particulates and nefarious bacteria.

Cystic fibrosis (CF) interferes with this cleansing process. A particular protein that sits inside the membranes of lung cells conducts chloride ions out of the cell, actively dumping them outside like bouncers after a bar fight. A mutated form of this gene, called cystic fibrosis transmembrane conductance regulator, fails to form active salt-transporting proteins; when a person has two copies of the mutated gene, the cells lining the lungs do not transport salt well. Because too little water rushes out of the cells to dilute the salt, in a cascade of implications, the mucus becomes too viscous to be moved by the cilia, the particulates and bacteria accumulate, and the lungs' housecleaning fails. The lung problems of CF— blockages, infections, decay of the broccolilike alveoli that farm oxygen out of the air—stem from a failure of the lungs to be self-cleaning, all traced to the failure of an otherwise anonymous protein to be a good salt stevedore.

Although we understand the gene well, the reason for so many CF mutations and their prevalence in some populations remains mysterious. Something must generate direct selection for CF mutations to explain their high frequency, but this reasoning swims upstream against the mutation's awful negative impact when a person inherits two copies. Because cystic fibrosis does not occur in people with only one copy of the mutated gene, perhaps the benefit of the CF mutation derives from its action while it is alone in the cell. In particular, if the CF gene protects its carriers somehow, then selection may favor it.

Is Cholera the Culprit?

Cholera kills in a messy disruption of salt balance. The bacteria that cause cholera secrete a toxin that has a dramatic effect on the human digestive system. Working its way into cells that line the intestine, cholera toxin activates the proteins that transport salt into the gut. Once activated, the floodgates of the cell open wide and salt rushes out, followed by a large amount of water. As much as five gallons of fluid a day may be transported into the gut, and voided by the poor sufferer of cholera. Water and salt loss causes dehydration, the typical killer in this disease.

Cholera toxin affects the cystic fibrosis salt transporter, and so people with a single CF mutation—instead of two—show a degree of protection from cholera's dehydrating effects. Cholera toxin still opens the floodgates, but only half the proteins respond. Perhaps, during cholera attack, only half the normal fluid and salt rushes out, and, because dehydration drives mortality from cholera, perhaps this blocking ability extends life long enough for people to rid themselves of the cholera bacteria.

This clever notion, proposed in 1994 by Sherif Gabriel and colleagues from the University of North Carolina, has gotten a shot in the arm by experiments on mice engineered to have human CF genes. Mice with one normal human salt transporter gene and a CF mutant gene—just like human carriers of the CF mutation—show less fluid loss when injected with cholera toxin than those with two normal human genes or two mouse genes. Although these experiments support the basic idea of an advantage of having one CF gene, the story remains crucially incomplete.

THE CYSTIC FIBROSIS gene shows up very commonly in Europe, but cholera historically has not been found there. The first cholera epidemics to reach Europe were in the early nineteenth century. Transported from India by the growing trade links to the Orient, cholera ricocheted around the cities of Europe and North America between the 1830s and 1896. The New York City epidemic of 1832 claimed thousands of lives, and 10 percent of the population of St. Louis died of cholera in 1849.

London water supplies spread cholera like the rail lines conducted trains, but the fight over what caused local outbreaks of cholera was only solved by a principled urban terrorist. In 1849, London was again wracked by cholera plagues, radiating in waves from mysterious centers to lay waste to whole neighborhoods, and physician John Snow pleaded his theory that the outbreak raged where the city's water supply was contaminated. Other medical theories prevailed until Snow took matters to the street, and removed the water pump handle on Broad Street, an epicenter of disease. Cholera died like the embers of a starved fire, cut off from constant renewal from the contaminated well; in this way the disease was tamed in that neighborhood.

Although cholera devastated Europe, its tardy historical appearance muddies our explanation, suggesting that cholera cannot have been the sole source of advantage for the CF mutation. Of course, cholera has a lot of terrible company, diseases that cause death from diarrhea and dehydration. Many other infections, even those caused by rogue strains of normally benign bacteria like *E. coli*, cause panic flushing of the digestive tract and huge fluid loss.

This entire suite of diseases becomes the probable historical culprit of selection on CF genes, raising their frequency in urban settings where digestive infections run riot due to close proximity and poor sanitation. In 1999, Gerald Pier and colleagues at the Brigham and Women's Hospital in Boston announced a potential second link between CF and bacterial disease. They showed that typhoid bacteria use the normal salt transporter protein as a clandestine door to help them enter mammalian digestive cells. In mice with one CF mutant gene and a normal salt transporter gene, these bacteria move across the gut wall only 14 percent as well as they do in mice with two normal genes. Like the work with cholera, these experiments tantalizingly hint about the role of the CF gene in the history of human disease.

Yet another problem greets this explanation: although urban diseases like cholera may have selected for the CF gene, such ailments have long afflicted urban centers outside Europe—those on the Indian subcontinent, for example. What causes the rarity of the CF gene in India? Likewise, diarrheal diseases span the globe, particularly on tropical continents, yet we find the CF gene in the tropics only rarely. Why?

Paul Quinton, a physiologist working at the University of California at Riverside, has proposed an additional link between cystic fibrosis and natural selection that may help to explain the geography of this disease. Salty sweat, he points out, has long been an early sign of cystic fibrosis. The CF protein acts inside one of three cell types in sweat glands, the beta-S cells, and alters their ability to regulate the salt concentration of sweat. Even people with just one CF gene lose more salt in sweat than people without the CF mutation. Quinton points out, "Until only a few hundred years ago, salt was a precious biological commodity." The salt trade across Europe, Africa, and Asia was immense: Gandhi's famous march to the sea was to break the back of the British-held monopoly on salt production in northwest India and return the means of salt production back to the local people. Because salt has been precious, losing too much salt every day in a hot climate might put people with the CF gene at a disadvantage over people without the mutation, a disadvantage that might outweigh the benefit of the CF mutation. Quinton speculates that this explains why the CF mutation has become common only in the cooler regions of Europe—cooler weather reduces daily salt loss, and in these circumstances, the selection for disease resistance outweighs the disadvantage in salt loss.

This explanation has its problems too. Europe can get very hot—Quinton has clearly not spent much time in Rome in summer—and cooler regions of northern China and North America have indigenous populations with low frequencies of CF mutations. Nevertheless, the link between sweat, lungs, guts, and the CF mutation shows how intricate the selective balance could be, as well as illustrating how a current human genetic disease may be a player in the drama of human evolution, past and present.

AIDS to Selection

HIV sneaks into white blood cells and sets up camp like an invading Bonaparte. Its violent evolution to thwart immune extirpation and to short-circuit drug destruction flares starkly during its reign as a kamikaze parasite. For most people, this evolution starts a desperate race that HIV will eventually win. But a few—one in 100 perhaps—can simply say this danger doesn't apply to them.

Moving through the community of AIDS sufferers has always been a small cadre of individuals exposed to the virus but uninfected by it, or infected but able to contain its spread. These are the lucky ones in whom HIV remains only a name, not a condition; they never develop full-blown AIDS. Orbited by anxious health researchers looking for the reason they can cut such an impervious swath through the sea of virus, they have been the raw material of hope for a decade, a hope rekindled by their unfailing health and persistent victory.

After doggedly exploring a thousand avenues where the reason for HIV tolerance might live, in the past few years optimistic researchers have discovered a remarkable fact about some of us. Some people have an ability to resist HIV naturally, and at least part of this resistance lies in our genes.

The reasons have been worked out in a series of improbable discoveries that have focused on a handful of genes out of our tens of thousands. The search required a decade's sleuthing into the role played by various proteins and hinged on the discovery of the locks that control access to white blood cells, locks that HIV must penetrate.

Cheetahs and Winners

Stephen O'Brien runs a diverse laboratory at the National Institutes of Health, and long has ferreted out the secret treaties between attack viruses and mammalian cancers. He lives a double professional life, delving into the genetics of human health crises and also working with the National Zoo on the genetics of endangered mammals. O'Brien thinks in a dynamic, evolutionary way. He see a disease and its host in evolutionary orbit around each other, and seeks to understand those orbits to know the nature of a disease and predict its future. O'Brien's obsession with genetic variability has taken him on many genetic safaris, and led him from cheetahs to mice and then to humans and HIV.

Cheetahs, the fastest cats, chase down some of the fastest groceries in the world. O'Brien's work in the 1970s and 1980s focused on a unique feature of cheetah populations—they so lack genetic variation that different individuals are practically clones of one another. For example, any cheetah can accept a skin transplant from any other. This feature differs startlingly from the complex testing that must precede a human transplant: unless we find exactly the right donor, a human skin graft will almost certainly fail.

Cheetahs lost this variable graft interaction when they lost their genetic diversity—probably through past episodes of very small population size and severe inbreeding. Without a wide variance in their gene pool, they are susceptible to epidemics and often mount poor immunological response to infectious diseases. By contrast, humans retain a great deal of variability, which makes a critical difference in disease response and contributes to the versatility of our immune systems.

O'Brien has tracked the world's mammals to squeeze their genetic diversity into his intricate measuring devices. Along the way he has dawdled to help invent the science of conservation genetics, helping to understand the connection between genetic diversity and genetic health. And with this perspective came a slowly simmering insight—perhaps the right mix of genetics could thwart a virus. Just as individuals differ in the mirror, perhaps they interact differently with viral invaders. Mice showed this early on. A population found outside Los Angeles was genetically resistant to a leukemia virus that rampaged through other local vermin. When the HIV epidemic began to rage through the world, O'Brien remembered the mice and began to wonder whether there was a gene that could block HIV, and thwart AIDS, hidden among the variegated leaves on the human genetic tree.

This was too important an idea to let slide. "We began a long, slow tedious process of collecting blood specimens from people at high risk of AIDS," O'Brien recounts. "Thousands of them. I was sure that there would be natural human differences which affected how people handled the AIDS virus." O'Brien's lab collected the samples and kept them safe.

Meanwhile, AIDS researchers were accumulating genetic information on long-term nonprogressors, as those who were infected with HIV but did not advance to the disease are called. Chasing down the reasons for this surprising resistance involved isolating white blood cells from resistant people, and testing their mettle by exposing them in the lab to some of the more voracious strains of HIV. In some cases, the stubborn white cells resisted infection during weeks of constant exposure to high virus levels.

Tantalizing evidence from around the world suggested the involvement of cellular hormones called chemokines; these are chemical signals that cause the cells of the immune system to activate and attack foreign invasions. Chemokines dock only at receptor proteins on white blood cells, and it turns out that the cells targeted for infection by HIV have such a docking receptor protein, which is known as CCR5. Even more crucial was a parallel discovery, made by researchers working on a totally independent line of inquiry about HIV function: they found that HIV uses the CCR5 protein as its second attachment site to get into CD4 cells.

HIV Always Knocks Twice

As we learned in chapter 5, HIV attaches to white blood cells with the CD4 protein on their exterior. But a virus attached only at this point—spinning like a circus performer suspended by her teeth over the distant floor—cannot invade the cell. Instead, the virus must attach twice—once to the CD4 protein and once to CCR5. Only when both locks open can HIV breach the cell membrane and slip inside. Without CCR5, cellular invasion is as easy as getting good grammar from a rock song—and without HIV infection there is no AIDS.

These results began to come together in early 1996, when researchers working on chemokines, resistant cells, and the HIV entry mechanisms suddenly saw a pattern in one another's discoveries and realized that CCR5 was a previously unknown but pivotal point in the HIV life cycle. A whirlwind surrounded these results. Within days of the announcement, O'Brien's lab looked for and found what they'd been waiting for—"A whopping large deletion mutation smack in the middle of the CCR5 gene" as O'Brien termed it. Not only was CCR5 variable in humans, the CCR5 gene that made a defective protein conferred HIV resistance. In all their samples, no one with two copies of the mutated CCR5 gene had HIV. Despite high-risk factors, all such individuals had escaped infection.

This resistant version of the CCR5 gene has a unique mutation eliminating a short string of amino acids in the protein, without which HIV cannot enter the white blood cell. Those who possess two copies of the mutant CCR5 gene resist normal HIV strains, although subsequent work has shown that an advanced HIV strain, ravaging through the exhausted immune system of an AIDS sufferer, can sometimes attack even these protected people.

Even a single mutated CCR5 gene (plus a normal one inherited from the other parent) results in fewer intact CCR5 protein molecules on the outside of CD4 cells, and slower entry of HIV into those cells. Although HIV infection can still occur, having one copy of the mutated CCR5 seems to slow HIV entry, resulting in slower disease progression and a longer time until HIV develops into full-blown AIDS.

Who has a mutated CCR5 gene? The roaring AIDS epidemic

suggested that the mutation was not in most people. O'Brien examined nearly 1700 blood samples from people with HIV. Yet another surprise greeted this effort—the CCR5 mutant surfaces very commonly, occurring in up to 20 percent of people, but only in some places. It occurs in one in five Europeans, but in less than one in one hundred Africans, native Americans, or East Asians.

Historic Selection

Estimates of the age of the CCR5 mutation placed its origin up to 50,000 years ago, a bit before modern humans began to replace Neanderthals throughout Europe. So, although currently at low frequency in Africa, the CCR5 mutation may have arisen there and thereafter spread into Europe. However, by itself this migration could not have increased the frequency of the CCR5 mutation. Natural selection provides enough power to increase CCR5 mutants, but what could generate the necessary selective pressure?

We could hypothesize that a prehistoric epidemic raged through Europe, and that the CCR5 deletion mutant conferred resistance on that plague as well. Could this epidemic have been a very early version of HIV, a separate incursion of the virus into the human stock? Or could it have been a different virus, one that latched onto white blood cells and entered through the CCR5 keyhole like HIV? Both ideas rank as pure speculation, but today, CCR5 selection can be understood on the basis of the homing mechanism of HIV, as well as seen as a contemporary force of evolution still active in the human population.

THIS SIMPLE DISCOVERY—that human genetic variation relates to AIDS development—galvanized the medical community because for the first time it was clear that HIV was not an overwhelmingly powerful foe. Further work revealed even more links between human genes and AIDS. In fact, we now know that a series of mutations in the CCR5 gene affects AIDS progression.

Reports soon appeared showing the effects on AIDS of variation in other human genes. The CCR2 gene has a mutation at position 641 that associates with lengthened progression to AIDS. This mutation helps reduce disease progression, even when found in only one of the two CCR2 genes. In a study of African women infected by HIV, mutations in CCR2 explained why some progressed to AIDS slower than others about half the time. A different gene, called SDF1, codes for the protein that attaches to a third HIV entryway to cells, a receptor called CXCR4. Overproduction of SDF1 gums up the CXCR4 receptor, blocks access to HIV, and reduces AIDS progression. The protective effect of SDF1 and the CCR5 and CCR2 genes appears additive—although having protective variants of all three genes occurs rarely, when these mutations do occur together, progression of HIV to AIDS slows dramatically. Two other human genetic loci—one called RANTES and the other responsible for some of the initial steps of immunological recognition—have been linked to either accelerated AIDS development or slower progression.

The year 1996 dawned without a single strong example of the impact of human genetic variation on AIDS. Now we know of a battery of gene sets that determine AIDS resistance in humans and can already explain about half the variation in progression time from person to person.

As the new century dawns, we suspect that other genes play a major role in HIV infection and AIDS development. This complexity will keep full understanding at bay for a long time. But the Pandora's box of human genetics has sprung open and we can begin to see how within the variation we possess as a species there may be answers to some of the problems that continue to plague us. The connection of our genetic variation to important health dilemmas represents the links forged by evolution, links that evolutionary thinking predicts and evolutionary optimists will attempt to unearth.

The Face of Human Evolution

A more complex terrain than the human face is difficult to imagine, but we seem remarkably able to recognize and remember it with a startling exactness. Our uniqueness extends deeper than facial features, deep into the intimate and hidden corners of our genetic prose, deep into the DNA sequences that beat molecular discipline into our cells and bodies and lives. Deep in those recesses, every human nurtures an idiosyncratic flame.

If we were any other species, then we would see natural selection stalking these divergent genetic skeins, picking and choosing, and like the piston of evolution it is, driving the evolutionary engine. But we like to think we miss traveling the evolutionary path by virtue of our religion or intelligence or technology, and so we tend to deny these things happen in our species. We like to think we have evolved beyond evolution.

Despite exclamations that human evolution ground to a static halt when our culture became so powerful, a pause to look at the varied conditions of the human community shows that, like any other species, we bow to natural selection. Even with our technology, we still face the varied tests of diverging life stresses, and when health, habitat, or nutrition are poor, then these stresses can kill. Our different faces matter most at this point—when they look out through the veil of desperate conditions and when some can see farther than others.

But even for these changes, human evolutionary shifts crawl slowly by the anxious time scales humans favor. We could be tempted to dismiss these shifts as unimportant window dressing, given our penchant for instantaneous results. After all, we are not going to evolve wings soon, no matter how bad the traffic. It might take 1000 generations—longer than the human race has known writing—for AIDS to shift the frequency of the CCR5 mutation from 10 percent to 90 percent. This seems slow to us because we are impatient beasts and 1000 generations seems too long a period to contemplate. This kind of genetic evolution in humans hasn't been slowed by our power to keep selection at bay, but instead, it merely seems slow as it proceeds, when we are in the middle of it.

Keeping us and our brethren safe and happy, independent of inborn genetic errors, glitters as a future goal. We hope to extend throughout the world the same power of technology that crafts the moist, thin crescents of my contact lenses and takes the sharp sugar edge off my milk. But our social compact does not bring all the advantages of expensive technology home to everyone, and the growing gulf between rich and poor on our planet puts 50 percent of the wealth in the hands of 5 percent of the population, while 1.3 billion people earn less than $1 a day. Although the benefits of modern medicine, civil sanitation, and high-calorie diets have been extended further and further into the world by dint of massive human effort, this new hope does not penetrate the worries of every life. The claim that evolution has completely stopped for humans is largely wishful thinking—a view that might be born in a sparkling maternity ward but not in a tin hut.

BARRING UNIVERSAL ACCESS to all the medical fixes our technology can offer, genetic evolution in humans will remain an active process. But another evolutionary mechanism churns in our species—human behavioral change surges and crashes like a wave on the beach of our culture. The tide of cultural change rises steadily, and we are constantly swept away by the waves of our own ingenuity, flooding last year's lifestyle. In the view of many psychologists and philosophers, among species, we are uniquely born to manage this cultural chaos. For many evolutionists, we are evolved to tolerate and even promote it. And for some students of the evolutionary process, it seems that our minds have created a new evolutionary track—one based on ideas, not genes, one based on communication, not inheritance. These students are the meme weavers, and they apply the principle of evolution to try to understand how our culture changes so quickly.

Chapter Ten

The Ecology and Evolution of Aloha

On Fridays, no matter what the weather, I show up to lecture in an aloha shirt. Boston winters make this sometimes an intrepid idiosyncrasy, but a thick-knit fisherman's sweater and a leather jacket keep me warm across the freezing Harvard courtyards, before I strip down to red flowers and blue fish in time to begin talking about evolution. I do this for two reasons. Like a good mood on a hanger, an aloha shirt on a cold, gray day preempts the winter doldrums and makes it much more difficult to maintain a December pout. But just as important swirls a simple, yet fascinating uncertainty: How well will the idea catch on?

At the beginning of each semester, while sunny fall days still promise rewards for eschewing tweed, I encourage undergraduate and graduate students to don aloha shirts too—all to inoculate an infectious good mood, and as an ongoing experiment in the ecology of ideas. Looking into a lecture hall full of new students, I can never predict the response. One year I had to go out and buy my teaching assistants aloha shirts to cajole them into such un-Boston colors, while in other years the idea spreads on its own and the lecture hall is a rainbow on Fridays.

All ideas arise in one brain and move into others on a carrier wave of language and example. Once in a new home, they dictate behavior—the aloha idea might surface in someone looking in a mirror on a cold Friday morning and take over for common sense—or they may fail to break an ingrained bias against hibiscus flowers and be ignored. Ideas change and sometimes improve once they spread to new brains—one year there were flower leis worn at lecture as well as aloha shirts—and they can take up what seems like an independent existence, or die the death of the forgotten.

These populations of ideas underpin our most powerful evolutionary invention—our culture—and rapid changes in the way we live our lives now come from cultural shifts rather than the kinds of physical alterations that have so far been the hallmark of evolution. What drives the extraordinary onslaught of cultural change? Can we find any similarity between the acceleration of change in society and the acceleration of evolution we encountered in previous chapters? Can evolutionary biology illuminate the human social environment, and if so, where does the light fall?

It turns out that the evolutionary engine can operate on ideas, but in a manner far different than it operates on finches or bacteria or salmon. The engine turns, but each of the three elements of variation, selection, and inheritance take on a new demeanor when applied to the evolution of ideas. Like a Picasso portrait, the elements combine in ways that challenge the limits of a flat canvas.

———

The Rapid Pace of Human Culture

Biologists often study evolution by comparing a trait across several generations—counting fly bristles or measuring guppy colors—and asking what changes, if any, appear. For observing different aspects of human evolution, my local shopping mall bustles as a laboratory, showing that cultural change outstrips physical evolution like fashion outstrips practical shoes. Differences between generations seem much more apparent in dress, demeanor, and even what body piercings we consider polite than physical features like brain size or upright posture. The former represent cultural changes that arose within the last few decades, showing up in major differences between generations, while the latter hark to past physical evolution.

Rapid cultural change ties together the very sinews of our recent history. The first known use of an alphabet for writing is found on a Middle Eastern rock in 1900 B.C., but within 1500 years alphabets were used to construct the compelling collection of plays, poetry, history, philosophy, and science that fueled the culture of Greece. Humans have wielded the written word for no more than 300 generations, not much longer than the number of generations evolutionists require to double the number of bristles on a fly thorax. But the changes in human lifestyles during these few millennia immeasurably exceed the simple, numeric changes in bristles recorded in experiments on fly evolution.

Whoever thought of yogurt probably just made the best of a bad thing. "Spoiled milk isn't all bad," she probably said. "It travels well and sticks to a spoon." But the idea of lacing milk with the right bacteria so it spoils correctly embodied a powerful and valuable tool, sweeping through Middle Eastern society. Innovations continue to sweep through the way we live, particularly accelerating the movement of information about the globe. The twentieth century opened its eyes to the genius of Edison recording scratchy words on a record, saw Marconi in England straining to punch a message to Newfoundland with feeble radio waves, and ended with a gigantic information network straining at the bounds of modern computer capacity. Such change affects every path our culture takes (the Dalai Lama has e-mail) and has come to define the essence of modern human society.

Fashion Slaves and Neanderthal Manners

Waves of fashion wash most rapidly up on our cultural shore. Signals of social rank and wealth, stylish modes of dress have come to serve a purpose similar to the garb of the resplendent male guppy—both are used to attract attention from potential mates or to subdue the optimism of rivals. Throughout Renaissance Europe, fashion began to obsess the wealthy. In *Hamlet*, Polonius advises his son, Laertes, as he departs for Paris:

> *Costly thy habit as thy purse can buy,*
> *But not express'd in fancy; rich, not gaudy:*
> *For the apparel oft proclaims the man.*

As Polonius knows, fashions were used as a social badge to identify the rich and to keep the middle class in a separate stratum. Fernand Braudel quotes a Sicilian traveler in 1714: "Nothing makes noble persons despise the guilded costume so much as seeing it on the bodies of the lowest men of the world." Throughout most of Europe, the poor had scant recourse to the fashion game played out in aristocratic courts. Braudel observes, "To be ignorant of fashion was the lot of the poor the world over."

Some cultural transitions start out slowly, and then catch on. Underwear became common among the rich in France by the thirteenth century, but it took until the eighteenth for all layers of society to follow suit. The invention of the individual fork—as opposed to the two-tined tool used to hold a gob of meat in place while carving it—occurred in about the sixteenth century but spread from Italy only slowly. Louis XIV wouldn't touch one—perhaps because he had already established a formidable reputation for being able to eat chicken stew neatly with his fingers—but the fork soon became the hallmark of polite society. And what we now call table manners required some legislative help to spread. In 1624, young Austrian officers, dining at the archduke's house, had to be required by ordinance not to arrive half drunk, spit on their plates, or wipe their noses on the tablecloth.

Such cultural shifts were rare in prehistoric societies. At the dawn of humanity, technical innovation of stone tools crept along. Roger Lewin writes that stone tool innovation was in an "incredibly slow phase leading from the earliest artifacts some 2.5 million years ago to approximately 250,000 years ago." Neanderthals from 250,000 to 50,000 years ago had a fairly static culture too. They had clothing but no needles; they lived in dwellings, but none were solid enough to survive to the present day. Their blandly utilitarian Mousterian stone tools included no advanced bone or antler implements. There were no boats. Artifacts were made only out of stone gathered from local sources, which indicates that there was no long-distance trade. And no art has survived.

So where did our current cultural maelstrom come from? When did our society develop a penchant for change? The genesis of the human ability to manipulate ideas and to thrive on the new capacities they provide remains largely shrouded in deep history. But along the steep slope toward becoming human, our ancestors probably developed a proficiency with ideas in association with advances in the information-exchange tool called language.

The Wardrobe of Thought

The ape wars of mid-nineteenth-century science pitted Thomas Huxley against Sir Richard Owen in a viciously fought battle over whether there were fundamental differences between humans and ape brains. Owen insisted the human brain had a unique organ called the hippocampus. Huxley knew very well that apes had such a structure too, and suspected Owen wanted only some thin excuse to demonstrate the uniqueness of humanity. In 1857, the London magazine *Punch* reported:

> *Says Owen, you can see*
> *The brain of a chimpanzee*
> *Is always exceedingly small,*
> *With hindermost "horn"*
> *Of extremity shorn*
> *And no hippocampus at all.*
>
> *Next Huxley replies*
> *That Owen he lies*
> *And garbles his Latin quotation.*
> *That his facts are not new*
> *His mistakes not a few*
> *Detrimental to his reputation.*

Although primate brains differ in proportions of the various sections, a twenty-first-century anatomist must agree with Huxley, and conclude that the basic structure of the ape and human brains mirror each other.

But overall size still matters. The human brain has grown steadily over the past 2.5 million years. Even in ancestors very similar to us in other ways, brain size measured only half our current volume as recently as 1 million years ago. Throughout the hunting and gathering era of the past Ice Ages, as glaciers counted the passing of 2 million years with a repetitive grinding migration across a continental stage, brain size increased slowly, a volume increase of about 5 milliliters (a quarter of a raindrop) per generation. But however slowly our brain grew, by about 150,000 years ago it had reached its pinnacle—the 1.2 liter engines we carry with us today and that we use to power our interactive society.

Did the cultural rigidity of our ancestors derive from their small brain size? Our fossil record suggests not, because Neanderthals had brains our size and larger, yet showed little advanced thinking and few flexible ideas. By contrast, fully modern humans, the society that replaced Neanderthals throughout Europe and the Middle East about 40,000 years ago possessed evocative art, complex tools, dwellings constructed with solid postholes, needles, boats, and almost certainly complex language.

What took so long? What sparked this cultural achievement? In *The Third Chimpanzee,* Jared Diamond floats the idea that a "Great Leap Forward," the acceleration of human cultural change and experimentation, occurred through the spark of advanced development of language as a medium for cultural exchange and innovation.

Although language itself counts as an idea, it has the great property of promoting and incubating other ideas, allowing transmission to others with greater accuracy. Before language, ideas jumped the gap from mind to mind through example.

But representational language made these jumps easier. At first, Helen Keller—her mind trapped by blindness and deafness, her early life set apart from normal language—could learn only what teachers could physically show her. Suddenly, in an insight that must have been just as difficult as the uttering of the first human spoken word, she invented for herself the idea of representational language, and her educational potential exploded.

So too, the human evolution of language provided a river for the flow of ideas from mind to mind. Language, the crucible of thought, the wardrobe of our ingenuity, creates a habitat that can hatch and remember ideas. Through language, human culture could accelerate, because through language the rate at which ideas could be transformed, tested, and transmitted increased to a furious pace—one faster than anything else in the history of life on Earth. Although ideas can move by example—seventeenth-century French fashions stormed through Europe on dolls outfitted so perfectly that dressmakers could copy the designs—ideas transmit much more easily when greased by words.

The Evolutionary Engine of Ideas

Some ideas, hardwired into animal brains and generally qualifying as instincts, are inherited like feather color or beak size. The songs of mating doves, different from species to species, and where female butterflies choose to lay their eggs, carefully rejecting all but the correct leaves for their offspring to chew, count as behaviors that offspring inherit from their parents. But many other behaviors, even in other animals, are culturally inherited through imitation, not genetically passed from one generation to the next.

Rosemary and Peter Grant at Princeton University documented among finches of the Galápagos Islands that different species on the same island have adopted different courtship songs. But occasionally, a female bird will drop an egg into a nest not her own, and sometimes this nest is home to the other species. The foreign egg hatches, and foster parents unknowingly raise the interloper. Of course, the baby bird grows up to look just like its biological parents, but it usually doesn't *sound* like them. Instead, males sing the courtship song of their foster parents, and females, when they mature, respond only to the song of their adopted species. This confusion results in mixed matings between species; cultural inheritance of courtship song, rather than strict genetic inheritance, scrambles the mate choice rules between two types of finches.

Like this birdsong, we humans pass on our entire cultural fabric from generation to generation without regard to genetics. This fabric underlies social interactions like a tablecloth lies beneath a dinner, absorbing cultural spills and covering over social splinters. Ideas make up the basic threads of this fabric, and we grow up wrapped in its folds.

The fabric passes from generation to generation through writing, language, and example, not by the transmission of chromosomes. However, the evolutionary engine does not require a particular mechanism of inheritance—it demands only that traits of parents tend to crop up again in their offspring. This cultural inheritance appears in some other animal species—like the humpback whales in which newborn calves follow their mother across the oceans for a year or two, learning a particular migration route. Whales have no written maps, can spy out no shortcuts, and change routes very seldom. Humans mastered the ability to encode information in words, increasing the accuracy and storage capacity of language by inventing writing. Like the strings of nucleotide bases in our DNA, strings of letters passed through the generations store much of our cultural inheritance.

The Fitness of Ideas

In the first episode of *Monty Python's Flying Circus*, Mr. Ernest Scribbler invents the world's best joke, a joke so funny that anyone who hears it dies laughing. Such a joke, a powerful and attractive idea, has no potential to spread because of strong selection against it: a joke that kills the listener is difficult to tell, even for John Cleese.

Ideas vary from individual to individual, duplicating and moving into new brains, where a powerful and unique form of selection operates on them. Bad ideas, rejected like anchovy daiquiris, live on only in a few people with fishy breath. Good ideas duplicate quickly and spread far and wide, generating clutches of mental ducklings, with some subsequently turning into brilliant swans and others fated to remain only brain geese.

Some ideas may spread because they advance reproductive success in the individuals that use them. The flea-borne Black Death stalked Europe in the fourteenth century, but the Jewish admonition against housing animals in the same place as people may have saved them from the worst depredations of this plague.

Other ideas have the opposite affect on fitness. The heavy-handed, heavy metal band Slipknot has gotten the idea that they should punch themselves in the face during performances, and so should their fans. They also hit one another with instruments, dive into drum sets, and leap off amplifiers. This creative idea has resulted in several hundred stitches (among the nine band members), two broken collarbones, a half dozen cracked ribs, spinal injuries, a partially severed finger, and enough cuts, scrapes, and bruises to employ an army of medics. Members of the audience beat one another up and trample those who fall in the mosh pit. Will this band catch on? Or will the idea expire with the band members?

Sometimes single ideas have little impact, but they combine to create powerful biological forces in our society. The easy detection of fetal gender, a strong preference for sons to inherit names and family fortunes, and reduced family size enforced in China and favored in Korea have combined to force a dramatic shift in sex ratio in these two countries. Usually boys outnumber girls at birth 51 percent to 49 percent, but now couples opt to give birth to male children instead of female children, bringing 108–110 boys into the world for every 100 girls in China and Korea. This gender skew promises to have a dramatic impact on marriage practices in a decade or so.

Some ideas without this kind of direct link to fitness (in the sense of survival or reproduction) can spread anyway, because they have the indirect effect of making individuals more comfortable or otherwise appeal to our senses. We choose the books we read, the movies we see, and, unless we have teenagers, the music we listen to, on the basis of what we like, not on the basis of our long-term fitness. Logic is not considered necessary by our society to the process of aesthetic choice.

In Barcelona, artists once tried to apply the nonlogic rule—where art objects don't have to be practical—to architecture. The sculptured constructions of the turn-of-the-century architect Antonio Gaudi contain the elements of well-working office buildings and homes, but the standard utilitarian features of these buildings were all given a nonlogical twist for style. Take a normal building, remove most straight lines, add some bright ceramic work and serpentine railings, then melt it slightly and wiggle the whole structure until parts of it twist and fancifully blend. Pull some rooflines up like peaks in whipped egg whites, and paint brightly. This is a Gaudi building, most spectacularly defined by the unfinished cathedral *Sagrada Familia* melting toward the sky a few blocks away from the old center of Barcelona. Proclaiming the triumph of the fanciful by blending art and construction, Gaudi buildings continue to stop tourist traffic in the modern city because of their breathless style.

However, Gaudi paid for his unique sense of design. Except for one wealthy patron, Gaudi's soaring art failed commercially because people tend to demand grim practicality of buildings, and they can't apply the nonlogic rule to architecture. He ended his career penniless—so exhausted emotionally and financially by the *Sagrada Familia* that when he died in 1926, hit by a passing bus, Gaudi was mistaken for a beggar.

Turning the Engine of Ideas

The variation of ideas, their transfer from brain to brain, and our selection among them qualify ideas as having all three factors necessary for operation of the evolutionary engine. Ideas that provide the brain they live in with an easier life are emulated by other brains, thus spreading the idea even further and ultimately creating a widespread cultural shift. But ideas do not have to increase individual fitness to become common: if they have an inflated ability to replicate and spread, or show strong persistence in individual brains, they will spread. Successful advertising jingles (for instance, "It's the real thing"), driven into our heads by incessant media repetition, require a conscious effort to dislodge. Certain theme songs ("A horse is a horse of course of course") are so insidious that we can't forget them without a hypnotist. Both these tactics for a successful idea are independent of their value in our lives.

The mutation, selection, and spread of ideas diversifies our culture, just like mutation, selection, and spread of species diversifies the world with 100,000 types of flowering plants and 1 million insects. Cultural change seems linked to this evolution of ideas. Some ideas thrive, giving rise to dense populations of offspring notions like writing and jazz music; others fail to persist, like chastity belts, thus dropping out of the idea village.

A shift in thinking has been engineered along the way here. Instead of discussing the evolution of humans and our culture, we now focus on the evolution of ideas separate from humans, resembling very much a discussion of the evolution of genes as entities separate from the evolution of individuals that carry them. In fact, Richard Dawkins, superb evolutionary communicator and teacher at Oxford University, identified just this type of possibility in *The Selfish Gene.*

By possessing all three required features of evolutionary change, ideas may evolve in their own right, spreading, changing, leaping from mind to mind, and not necessarily doing the brains they inhabit any special good. Dawkins considered imitation the basis of idea replication, and named the replicating units, composed of memory and imagination, "memes." Memes are ideas, tunes, inventions, retorts, ways of doing business, ways of asking for help, and ways of saying hello. Because those memes that best duplicate and spread to other brains become cultural icons, they might evolve through a process akin to the action of the evolutionary engine. If so, then insights about the evolutionary process may help us understand why ideas sweep through our culture so rapidly.

This notion, that ideas evolve by a process akin to natural selection—and that this in turn explains the rapidity of human cultural change—has been widely discussed by evolutionists and psychologists. Tempers run hot on this subject, some influential writers accepting the basic premise lock, stock, and barrel, others disdaining the entire premise. Stephen Jay Gould writes in *Full House* about human cultural change, "The common designation of 'evolution' then leads to one of the most frequent and portentous errors in our analysis of human life and history." In fact, most insightful analyses of the evolution of ideas share at least this one feature—they may entertain the notion that Darwinian principles act on ideas—but like a flashlight with dying batteries, it casts only a wan light on the subject. In *How the Mind Works*, MIT's Steven Pinker, at odds with Gould about a lot of things, nods his head on this: "I think you'll agree that this is not how cultural change works."

Not academic backpedaling, these arguments attempt to find the limits of insight provided by an evolutionary perspective. We can learn some things about culture by slaving it to the Darwinian process, but human culture's rapid pace cannot be entirely explained by simple evolutionary metaphors. Memes and genes resemble one another only to a point, revealing fundamental differences on close scrutiny. In particular, because of the basic link between the ecology of organisms and how selection works them over, and because the ecology of ideas and organisms differ so, then the way selection acts on ideas will differ markedly from how it acts on critters.

The Ecology of Ideas

We can fit only a certain number of events into our family's weekly schedule, but my daughter can usually plan a list of activities that would kill us all. Planning, or course, is easy. Being at school an hour early, making muffins for the Dungeons and Dragons club, staying for orchestra or chorus, and trying out for the musical before riding lessons and playing at a friend's . . . all this, although easy for her to imagine, is impossible to do.

Because ideas have no width, length, or height, no need for real space, the ecology of ideas differs from the ecology of any living thing. Ideas have no carrying capacity, no number too big for a time slot, brain, or society. The brain of a creative child is flashing and crisscrossing with thoughts bouncing around like hot atoms in a flame. Her head may feel ready to burst, but it won't, because the ideas weigh nothing, require no bunk space, and don't mind the crowding.

This contradicts some other viewpoints about the packing limits of the mind. David Star Jordan, a famous Stanford ichthyologist, once said, "For every person's name I remember I forget the name of a fish." In *Darwin's Dangerous Idea*, Daniel Dennett writes, "Each mind has a limited capacity for memes," and uses this idea to promote the competition of memes for entry into brains. Like competition to get into the best car at the bumper car ride at a carnival, he proposes that ideas compete for access to brains and that, once filled, brains cannot hold more.

Of course, a single brain cannot hold and access all the world's ideas, so Dennett must be correct at some level. But the human brain has no recorded maximum capacity for ideas. A brain might tire of absorbing ideas, but even after a grueling session reading population genetic theories or tax form instructions, any brain can shift focus and absorb the starting roster for a basketball game or scan a menu for the right soup.

If ideas have no occupancy limit in a brain, they do not compete with one another for space in our mental habitat and can crowd within a mind in infinite numbers. True, some may make way for others in our upper consciousness—the only real way to drive the theme song from *Gilligan's Island* out of your head is to displace it by humming the theme from *Mr. Ed*—and this competition for active attention from the brain, rather than absolute competition for brain space, may be what Dennett means. I would agree with this wholeheartedly, especially as I drive frantically from orchestra to riding lessons with my daughter. Nevertheless, the exhortation of Malthus—that populations of organisms always grow to exhaust their resource base—fails for populations of weightless ideas crowding into human brains.

NOT ONLY DO ideas occupy space differently, they spread differently too. When we tell someone a new lawyer joke, it doesn't leave our heads just because it enters someone else's. We hang on to it until we find somebody else who hasn't heard the one about the Doberman, because for ideas, transfer and replication are the same. The idea, cloned and passed on, perhaps changing a bit in the process, grows in copy number like a bacterium replicating in culture broth.

But idea reproduction and replication of bacteria are not the same. The duplication and spread of ideas can happen with lightning speed, especially as information conduits increase in speed and breadth. As media outlets have grown in extent and complexity, the duplication rate of ideas has skyrocketed. Printing with movable type, introduced in Europe by Gutenberg in Mainz, Germany, spread slowly through Europe starting in the 1440s. Fifty years later, print shops in 236 European towns had turned out some 20 million books. This was a revolution at the time, but by today's standards we would consider this rate of idea spread monstrously slow.

With no limit on reproduction rate, the concept of reproductive fitness—how well an individual fly or gene makes copies of itself—quickly loses meaning. We can calculate such a thing for a bacterium dividing: if every cell doubles each hour, then starting from a single cell, we count 32 after five hours, and in two days about 300,000,000,000,000 cells would swarm. How would we calculate this for ideas? They do not replicate at any regular rate—an idea may find a new home in a billion brains during a Super Bowl commercial, and then never replicate again. Because of this incredible uncertainty, we can't label an idea with a future rate of spread like a bacterium, and so we can't predict the number of copies of an idea in a day or a week. Instead, we can only look back to see how an idea has grown, documenting how many people have heard of a particular thing by a particular time.

IDEAS ARE BORN in a cauldron of imagination, but how do they die? If they have not jumped to other brains beforehand, they will die with the brain that houses them. Evidence of Fermat's last theorem existed only as a cryptic note scribbled on a page of mathematics results in 1637. "I've found a remarkable proof of this fact," Pierre de Fermat wrote, "but there is not enough space in the margin to write it." And so the solution to a famous math puzzle died with the author and waited until 1993, when mathematician Andrew Wiles announced the answer.

But we rarely see the demise of an idea in such a clear-cut manner. Instead, the human brain has the keen ability to resurrect a dead idea, dust it off, and let it live again. Like wearing bell-bottom pants long stored in the back of the closet, the human mind can easily don an idea previously rejected. Because memes do not die like real organisms, we can always call them back into service. Resurrection, only a casual thought away, can be easily sparked by an evocative smell or the tint of an old photo. Cavalry remains a widely recognized idea, even though no longer a winning military strategy. Other ideas survive their fame like child film stars, waiting for the chance to reappear or fading into the cultural background. We resurrect ideas so easily because new memes just pile on top of old ones in our prodigious memories, burying them in deep mental recesses but seldom destroying them.

Why Ideas Evolve Differently

A while back, somebody thought of white-out, the white fluid that you dab over a typing error, allowing you to retype the correct letter—a sort of Neanderthal spell checker. After somebody noticed that white is not the only paper color used in the world, and particularly that invoices and bills often come in blue and pink, then "mutant" white-outs in blue and pink were born. I wasn't there, but I'll bet that pink white-out didn't happen by a whole range of random, pale colors being produced and tested against a range of different papers to see which ones were the most successful. Mutation and natural selection would generate different color moths this way, and choose the most successful among them, but it isn't how human invention works. The common colors of paper used in offices were probably determined, and the white-out color adjusted to match.

This example shows that ideas do not change by some random mutation—a big difference between the evolution of the Beatles and the evolution of beetles. Some ideas, especially rumors, get warped in transmission, but most change by conscious improvement, the human act of thoughtful editing and careful addition.

We give birth to ideas in at least one other way—by uncontrollable macromutation. Our idea machines, not under perfect command, spit out unbidden notions in our imagination. Sometimes our thoughts go where they will, often producing radical departures from the parental thoughts that spawned them. We often call these departures "imagination," and when coupled with conscious selection, imaginative ideas can lead to the innovations that fuel rapid cultural shifts. This represents another major difference between idea mutation and gene mutation: genetic mutations usually occur through relatively small changes. But for memes, mental macromutations and the monsters they create abound. The creative process generates major innovations sometimes by uncontrollable happenstance and random brain action.

IN *ANNIE HALL*, Woody Allen and Diane Keaton conduct a common dating ritual on a New York rooftop—casual conversation with a mildly panicked need to appear witty. Their spoken conversation, inane dating chat that they both disdain, overlies a very different, unspoken dialogue—about each other and about themselves—visible on the movie screen. But the alternative lines are kept out of the character's mouths by force of will, an example of conscious selection for social self-preservation.

The sifting and sorting of ideas, the logical ones and the ones unbidden, by a process of conscious selection makes up the next stage in the evolutionary drama of the meme. Conscious selection works hand in glove with conscious improvement of ideas—like pink white-out—exhibiting a creative aspect missing in patterns of natural selection, a way of making a silk purse out of a sow's ear where natural selection would continue to try to make a better ear.

One of the biggest differences between the conscious selection and natural selection of ideas arises from the ability of conscious selection to predict the value of an idea before natural selection has had a chance to work on it. Bungee jumping works as an example, since a successful jump depends entirely on whether your cord is shorter than the height of the bridge you jump off. Conscious selection can evaluate whether a jump is a good idea given length of cord and height of bridge. In the recorded cases where this prediction has been ignored, empirical evidence shows that natural selection for arithmetic skills is alive in the world. If one hundred people show up at a 100-foot jump with cords ranging from 50 to 150 feet, natural selection would send them all over the guardrail and wait to see who shows up at the barbecue afterward. Conscious selection lets people figure out ahead of time who should jump and who should demure.

Only conscious selection allows the estimation, prediction, and evaluation of potential selective consequences, leading to a shortening of an overlong bungee cord or the search for a higher bridge. But a further insight emerges here: the heart of the human creative process combines the intuitive alteration of ideas with their conscious selection, effectively blending the mutation and selection aspects of idea evolution. Dan Dennett dwells on this combination without naming it in *Darwin's Dangerous Idea*, in an example of J. S. Bach's intuitive music. How does anyone compose a song? I compose mine by idle tinkering until something sounds good, a crude but direct imitation of the mutation-selection cycle that gives us sex change in the natural world and songs called "Sex Change" in the cultural one. But Bach composed his as a master who need not randomly pluck out a melody and try it out on his friends. Bach composed during a simultaneous mutation-selection firestorm of creativity to produce perfectly harmonious sounds. Other brilliant composers follow this model: Beethoven was nearly fully deaf by the time he wrote his Ninth Symphony. For Beethoven to sit and imagine such great and powerful music, without even being able to hear most of it, represents a merger of the mutation and selection elements of the evolutionary engine like a black hole collapses the elements of a star.

THE TRANSMISSION OF ideas from generation to generation follows the river of language, rather than the rigid train tracks of DNA. However, by itself, this does not constitute a big difference in how ideas and genes evolve. Cultural transmission of ideas takes place in the fluid world of humpback whales, or songbirds learning from their fathers to favor a particular song. But these different inheritance pathways do not lead to remarkable innovations of whale migration routes or bird repertoire. In humans, idea transmission and cultural inheritance have become different than what we see in most other animals. In humans, conscious selection interferes with what ideas are inherited.

With some exceptions, like what tune runs through my head early in the morning, I can usually choose the ideas I accept and pass on. Given a hundred good ideas in a book, I can grab just a few; given a dozen memorable scenes in a movie, I will remember only a handful in crisp detail. And this sifting of ideas during transmission, which is based on some judgment about how well those ideas will serve me in the future, contrasts sharply with the blind inheritance of genetic variants dictated by Mendel's rules.

Dress designs at the Academy Awards may provide a good example. What works and what doesn't appears clear in the reactions of the Academy audience, and in the following week's issue of *People* magazine. The good ideas live on the next year, and the bad ones resurface less often. Nobody has to die for this to happen (although rumors of links to reproductive success fly in the tabloids); instead, conscious selection acts at the stage of idea transmission to favor some and discard others.

THE IMPACT OF conscious selection at the stage of idea mutation and transmission blurs the distinction among the three elements of Darwin's engine and suggests a very different way of looking at ideas than Dawkins's notion of evolving memes. Picked over as carefully as meatballs at a cheap buffet, ideas are sorted by the finicky process of conscious selection. They are created, used, and discarded by active minds seeking their own advancement or their own comfort. What other element of our lives do we consciously improve for better function and pick carefully among to fill our cultural shopping carts? We can also consider ideas as tools.

As tools, ideas may be practical or not. They may have general or specific uses. Others may scorn them or adopt them with gusto. Sometimes they seem to have a life and independence of their own, like the wooden handle of an ax that becomes polished through use to fit the hands that wield it. But in the final analysis they remain tools, ways of manipulating the world or understanding it. They do not evolve like genes because like tools, they cannot really replicate themselves—they can be made only on demand by brains, and only by this agency can they spread through to other brains. This does not say they always benefit us—akin to the way many of us have toolboxes stuffed full of things not currently doing us any good—and it does not claim that they can never do damage—like an unchaperoned gun. But the function and rapid change of ideas does not require their independent evolution—it just requires the action of the co-opted evolutionary forces, directed variation, and conscious, predictive selection.

———

Accidental Evolution

Humans have a great impact on evolution: that has been the major point of this book, one that wraps up the reason for understanding the biological process of evolution in a practical package and hopefully delivers it home. Our impact on bacteria, viruses, insects, and weeds remains largely accidental. We do not intend that they evolve a resistance to our chemical tools, nor do we intend to generate new abilities in the species around us; nevertheless, our actions have generated a burst of evolutionary change that affects the entire natural world. And these changes cycle back to haunt us, affecting our daily lives.

Also accidental is our evolutionary impact on ourselves—the increased incidence of disease transmission among people living in dense urban settings, or the indirect effect of poverty on the ability to fend off infectious assaults. Finally, we can easily see our effect on our own cultural evolution.

But the evolution of ideas owes its frantic burn to elements that differ strongly from the normal operation of the evolutionary engine. Conscious selection, in its ability to predict the value of an idea, resurrect a long abandoned theory, or choose among competing ideas in different circumstances, is a fleet cousin to the plodding of regular natural selection. Even the generation of new variants differs startlingly in cultural evolution: the intensity of creative synthesis, the clever juxtaposition of divergent thoughts, and the pure leap of intuition or intellect that accompanies innovation count as critical features of cultural tsunamis that have only uncommon parallels in natural selection and mutation. Yet, the evolutionary engine has still been of great service in exposing the differences between cultural and organic evolution in a way that allows the operation of basic evolutionary principles while at the same time highlighting the unique aspect that conscious thought brings to the evolutionary stage.

WALK OUT OF your door and find some evolution. If you head for a hospital instead of the Galápagos Islands, I'll find my task accomplished. If you call your school board to protest their dismissal of the science of evolution from the curriculum by pointing to the herbicide-resistant weeds in the football field, then my writing has been rewarded. If you look in the mirror on a Friday and you're wearing an aloha shirt for no good reason, then that's the best compliment of all.

Sources and Suggested Reading

Chapter 1: From the Mountains to the Sea

For more about Hawaiian natural history, see E. A. Kay, *A Natural History of the Hawaiian Islands: Selected Readings*, vol 2. (Honolulu: University of Hawaii Press, c. 1994).

Chapter 2: Right Before Your Eyes

p. 10   Musings about artificial selection occupy a central place in Darwin's original arguments in C. Darwin, *On the Origin of Species* (London: John Murray, c. 1859).

p. 11   For a very readable exploration of the history of Darwinism, see L. Eiseley, *Darwin's Century: Evolution and the Men Who Discovered It* (New York: Doubleday, 1958).

p. 14   Early experiments on laboratory evolution of fly bristles are described in E. C. MacDowell, "Bristle Inheritance in *Drosophila*. II. Selection," *The Journal of Experimental Zoology* 23 (1915): 109–146.

p. 18   For a solid discussion of the experiment on the evolution of mouse body size and quantitative genetics in general, see D. S. Falconer, *Introduction to Quantitative Genetics* (New York: John Wiley, 1981).

p. 19   Experiments in fox sociality are described in D. K. Belyaev, "Destabilizing Selection as a Factor in Domestication," *Journal of Heredity* 70 (1979): 301–8. A more general summary is in R. Coppinger and M. Feinstein, "Hark! Hark! The Dogs Do Bark— and Bark and Bark," *Smithsonian* 21:119–29.

p. 21   For a description of T-shirt-smelling experiments and their relevance to human preference of partners, see C. Wedekind et al., "MHC-Dependent Mate Preferences in Humans," Proceedings of the Royal Society of London, Series B, *Biological Sciences* 260 (1995): 245–49.

p. 22   For a summary of laboratory approaches to producing species, especially in flies, see W. R. Rice and E. E. Hostert, "Laboratory Experiments on Speciation: What Have We Learned in Forty Years?" *Evolution* 47 (1993): 1637–53.

p. 25   Anne Houde's book reviews sexual selection in guppies, including citations to many seminal papers by John Endler, David Reznick, and coworkers: A. Houde, *Sex, Color and Mate Choice in Guppies* (Princeton, N.J.: Princeton University Press, 1997).

p. 28   Evidence of rapid diversification of sparrows after their introduction in North America is presented in R. F. Johnson and R. K. Selander, "House Sparrows: Rapid Evolution of Races in North America," *Science* 144 (1964): 548–60.

p. 29   The record of the results of the famous sparrow fall and measurements of who lived and died is contained in H. C. Bumpus, "The Elimination of the Unfit as Illustrated by the Introduced Sparrow, *Passer domesticus*," Biological Lectures, Woods Hole Marine Biology Labs, 1898, pp. 209–26.

p. 32   A Pulitzer Prize–winning book highlighting the work of Rosemary and Peter Grant on the evolution in Darwin's finches in the Galápagos Islands is Jonathan Weiner, *The Beak of the Finch: A Story of Evolution in Our Time* (New York: Knopf, 1994). It is a great introduction to Galápagos habitats, birds, and scientists.

p. 33   The Grants have published many papers and books of their own, including the authoritative Peter R. Grant, *Ecology and Evolution of Darwin's Finches* (Princeton, N.J.: Princeton University Press, 1999).

p. 35   "was they made, or only just happened?": Mark Twain, *The*

Adventures of Huckleberry Finn (New York: Grosset & Dunlap, 1918), ch. 19.

Chapter 3: The Engine of Evolution

Engaging books about the evolutionary process include G. C. Williams, *Natural Selection: Domains, Levels and Challenges* (New York: Oxford University Press, 1992). The development of Darwinism is chronicled by Loren Eiseley in *Darwin's Century: Evolution and the Men Who Discovered It* (New York: Doubleday, 1958).

p. 40   Scott Baker's exploits in Alaska are reported in: C. S. Baker et al., "Migratory Movement and Population Structure of Humpback Whales (*Megaptera novaeangliae*) in the Central and Eastern North Pacific," *Marine Ecology Progress Series* 31 (1986): 105–19.

p. 42   Anolis lizard ecology and evolution in the Caribbean are described in J. Roughgarden, *Anolis Lizards of the Caribbean: Ecology, Evolution and Plate Tectonics* (New York: Oxford University Press, 1995), and J. B. Losos, K. I. Warheit, and T. W. Schoener, "Adaptive Differentiation Following Experimental Island Colonization in Anolis Lizards," *Nature* 387 (1997): 70–73.

p. 44   *Horton Hatches the Egg* is by Dr. Seuss (New York: Random House, 1976).

p. 45   The European magpies and their egg-heaving ways are described in a paper by J. J. Soler et al., "Genetics and Geographic Variation in Rejection Behavior of Cuckoo Eggs by European Magpie Populations: An Experimental Test of Rejector-Gene Flow," *Evolution* 53 (1999): 947–56.

p. 46   The village weaver story is in A. Cruz and J. W. Wiley, "The Decline of an Adaptation in the Absence of a Presumed Selection Pressure," *Evolution* 43 (1989): 55–62.

p. 50   Hermon Bumpus's 1898 study: H. C. Bumpus, "The Elimination of the Unfit as Illustrated by the Introduced Sparrow, *Passer domesticus*," Biological Lectures, Woods Hole Marine Biology Labs, 1898, pp. 209–26.

p. 50   The evolution of reproductive strategies in the side-blotched lizard is described by papers by Barry Sinervo and friends: B. Sinervo, "The Effect of Offspring Size on Physiology and Life-History," *BioScience* 43 (1993): 210–18; and B. Sinervo and D. F. DeNardo, "Costs of Reproduction in the Wild: Path Analysis of Natural Selection and Experimental Tests of Causation," *Evolution* 50 (1996): 1299–313.

p. 58   The goldenrod gall story has been extensively studied by Warren Abrahamson and Arthur E. Weis; see their *Evolutionary Ecology Across Three Trophic Levels: Goldenrods, Gallmakers, and Natural Enemies* (Princeton, N.J.: Princeton University Press, 1997).

p. 63   The relationship between Darwin, Wallace, and Edward Blyth can be found in L. Eiseley, *Darwin and the Mysterious Mr. X: New Light on the Evolutionists* (New York: Dutton, 1979).

Chapter 4: Temporary Miracles

Overall suggested readings about antibiotic resistance include S. B. Levy, *The Antibiotic Paradox: How Miracle Drugs Are Destroying the Miracle* (New York: Plenum Press, 1992).

p. 67   For hyssop and antibiotics, see F. Bianchini and F. Corbetta, *Health Plants of the World: Atlas of Medicinal Plants* (New York: Newsweek Books, 1975).

p. 67   Some of the history of the discovery of antibiotics can be found in G. MacFarlane, *Alexander Fleming: The Man and the Myth* (Cambridge, Mass.: Harvard University Press, 1984), and R. W. Clark, *The Life of Ernst Chain* (New York: St. Martin's, 1985).

p. 69   Legends about the Florey and Heatley flight to Lisbon were discussed by Tom and Ray the Tappet Bros. on *Car Talk*, National Public Radio, February 1999.

p. 72   Zyvox approval and cure rates have been reported recently in "First Antibiotic in New Class of Drugs Fights Resistant Infections," *FDA Consumer* 34:4.

p. 75   Table 4:1: Levy, *Antibiotic Paradox*; Ciba Foundation Symposium 207, *Antibiotic Resistance: Origins, Evolution, Selection and Spread* (West Sussex, U.K.: John Wiley, 1997); C. Amabile-Cuevas,

Molecular Biology Intelligence Unit: Origin, Evolution and Spread of Antibiotic Resistance Genes (New York: CRC Press, 1993); and B. P. Rosen and S. Mobashery, *Resolving the Antibiotic Paradox: Progress in Understanding Drug Resistance and Development of New Antibiotics* (New York: Plenum Press, 1997).

p. 78   Data on the bacteria in the Murray Collection were reported by V. M. Hughes and N. Datta, "Conjugative Plasmids in Bacteria of the 'Pre-antibiotic' Era," *Nature* 302 (1983): 725–26.

p. 81   Transfer of vancomycin resistance was reported by W. C. Noble, Z. Virani, and R. G. A. Cree, "Co-transfer of Vancomycin and Other Resistance Genes from *Enterococcus faecalis* NCTC 12201 to *Staphylococcus aureus*," *Microbiology Letters* 93 (1992): 195–98.

p. 83   "the rate of spontaneous mutation . . .": D. E. Snider and K. G. Castro, "The Global Threat of Drug-Resistant Tuberculosis," editorial in *New England Journal of Medicine* 338 (1998): 1689–90.

p. 84   The relapse rate in TB as function of drug exposure was reported by R. Lewis, "The Rise of Antibiotic-Resistant Infections," *FDA Consumer*, September 1995, pp. 11–15.

p. 85   Resistant TB was described for a general audience by M. H. Cooper, "Combating Infectious Disease," *CQ Researcher* 5 (June 9, 1995): 489–96.

p. 85   Costs of antibiotic resistance are estimated in J. J. Schentag et al., "Infection Control and Changes in the Antibiotic Formulary for Management of Epidemic and Endemic Vancomycin-Resistant *Enterococcos faecieum*," *Hosptial Practice Special Report*, October 1998, pp. 22–36.

p. 87   Involuntary detention to force TB cures was reviewed by R. Gasner et al., "The Use of Legal Action in New York City to Ensure Treatment of Tuberculosis," *New England Journal of Medicine* 340 (1999): 359–66, as well as B. H. Lerner: *Contagion and Confinement: Controlling Tuberculosis Along the Skid Road* (Baltimore: Johns Hopkins University Press, 1998).

p. 90   Hospitals in Greece: M. Arvanitidou et al., "Antimicrobial Resistance and R-factor Transfer of Salmonellae Isolated from Chicken Carcasses in Greek Hospitals," *International Journal of Food Microbiology* 40 (1998): 197–201.

p. 92 For bacterial evolution and the cost of resistance, see S. J. Schrag, V. Perrot, and B. R. Levin, "Adaptation to the Fitness Costs of Antibiotic Resistance in *Escherichia coli*," Proceedings of the Royal Society of London, Series B, *Biological Sciences* 264 (1997): 1287–91.

p. 93 For bacterial replacement therapy, see G. Martin et al., "Inhibition Phenomena Between Salmonella Strains—A New Aspect of Salmonella Infection Control in Poultry," *Deutsche Tieraerztliche Wochenschrift* 103 (1996): 468–72; and U. Methner et al., "Combination of Vaccination and Competitive Exclusion to Prevent Salmonella Colonization in Chickens," *International Journal of Food Microbiology* 49 (1999): 35–42.

Chapter 5: The Evolution of HIV

Chapter 1 of Scott Freeman and Jon C. Herron, *Evolutionary Analysis* (Englewood Cliffs, N.J.: Prentice Hall, c. 1998), provides an excellent overview on HIV evolution. Also see K. A. Crandall, ed., *HIV Evolution* (Baltimore: Johns Hopkins University Press, 1999).

p. 100 Chimpanzee ancestor of HIV was described by F. Gao et al., "Origin of HIV-1 in the Chimpanzee *Pan troglodytes troglodytes*," *Nature* 397 (1999): 436–41.

p. 101 The HIV sequence from a 1959 blood sample was analyzed by Tuofu Zhu et al., "An African HIV-1 Sequence from 1959 and Implications for the Origin of the Epidemic," *Nature* 391 (1998): 594–97.

p. 102 Type C selection for heterosexual transmittion is reported by L. H. Ping et al., "Characterization of V3 Sequence Heterogeneity in Subtype C Human Immunodeficiency Virus Type 1 Isolates from Malawi: Underrepresentation of X4 Variants," *Journal of Virology* 73 (1999): 6271–81.

p. 103 Figure 5.1 is redrawn from S. A. Seibert et al., "Natural Selection on the gag, pol and env Genes of Human Immunodeficiency Virus 1 (HIV-1)," *Molecular Biology and Evolution* 12 (1995): 803–13.

p. 107 For the evolutionary response to the immune system, see S. M. Wolinsky et al., "Adaptive Evolution of Human Immunodeficiency Virus-Type 1 During the Natural Course of Infection," *Science* 272 (1996): 537–42; and Y. Yamaguchi and T. Gojobori, "Evolutionary

Mechanisms and Population Dynamics of the Third Variable Envelope Region of HIV Within Single Hosts," *Proceedings of the National Academy of Science (USA)* 94 (1997): 1264–69. Figure 5.3 is redrawn from the latter source.

p. 118 AZT resistance within five to twelve months: B. A. Larder and S. D. Kemp, "Multiple Mutations of HIV-1 Reverse Transcriptase Confer High-Level Resistance to Zidovudine (AZT)," *Science* 246 (1989): 1155–58.

p. 119 Resistance to ddI evolved within twelve months: M. H. St. Clair et al., "Resistance to ddI and Sensitivity to AZT Induced by a Mutation in HIV-1 Reverse Transcriptase," *Science* 253 (1991): 1557–59.

p. 121 AZT resistance: D. R. Kuritzkes et al., "Drug Resistance and Virologic Response in NUCA 3001, a Randomized Trial of Lamivudine (3TC) Versus Zidovudine (ZDV) Versus ZDC plus 3TC in Previously Untreated Patients," *AIDS* 10 (1996): 975–81.

p. 121 The study that showed that 3TC mutation increased reverse transcriptase fidelity is M. Wainberg et al., "Enhanced Fidelity of 3TC-Selected Mutant HIV-1 Reverse Transcriptase," *Science* 271 (1996): 1282–85.

p. 122 For indinavir resistance, see J. H. Condra et al., "Genetic Correlates of *in vivo* Viral Resistance to Indinavir, a Human Immunodeficiency Virus Type 1 Protease Inhibitor," *Journal of Virology* 70 (1996): 8270–76.

p. 122 Original protease functioned up to thirteen times faster: S. Gulnik et al., "Kinetic Characterization and Cross-resistance Patterns of HIV-1 Protease Mutants Selected Under Drug Pressure," *Biochemistry* 34 (1995): 9282–87.

p. 124 The three *Star Trek* episodes are "And the Children Shall Lead" (attack of a friendly angel), "Operation: Annihilate" (flying neurons), and "The Man Trap" (a salt-sucking alien).

p. 125 A hidden source of HIV: J. Wong et al., "Recovery of HIV-Competent HIV Despite Prolonged Suppression of Plasma Viremia," *Science* 278 (1997): 1291–95. However, the fact that drug cessation can speed the evolution of resistant or partially resistant strains was also shown by the same team in J. Wong et al., "Reduction of HIV-1 in Blood and Lymph Nodes Following Potent Antiretroviral Therapy and the Virologic Correlates of

Treatment Failure," *Proceedings of the National Academy of Science (USA)* 94 (1997): 12574–79. See also H. Gunthard et al., "Evolution of Envelope Sequences of Human Immunodeficiency Virus Type 1 in Cellular Reservoirs in the Setting of Potent Antiviral Therapy," *Journal of Virology* 73 (1999): 9404–12.

p. 126 "Patients previously treated . . .": R. M. Gulick et al., "Treatment with Indinavir, Zidovudine, and Lamivudine in Adults with Human Immunodeficiency Virus Infection and Prior Antiretroviral Therapy," *New England Journal of Medicine* 337 (1997): 738.

Chapter 6: Poisoning Insects, and What They Can Do About It

General readings include Mark L. Winston, *Nature Wars: People Versus Pests* (Cambridge, Mass.: Harvard University Press, 1997); Robert S. Desowitz, *The Malaria Capers: More Tales of Parasites and People, Research and Reality* (New York: Norton, 1993); D. N. Alstad and D. A. Andow, "Managing the Evolution of Insect Resistance to Transgenic Plants," *Science* 268 (1995): 1894–96; and G. P. Georghiou, "The Magnitude of the Resistance Problem," in *Pesticide Resistance: Strategies and Tactics for Management,* ed. National Research Council (Washington, D.C.: National Academy Press, 1986).

p. 134 DDT and the history of its use was described by T. R. Dunlap, *DDT: Scientists, Citizens and Public Policy* (Princeton, N.J.: Princeton University Press, 1981).

p. 135 For DDT use and abuse, see J. R. Busvine, "DDT: Fifty Years for Good or Ill," *Pesticide Outlook* 1(1989): 4.

p. 137 Part of the war on malaria is chronicled in L. Garrett, *The Coming Plague: Newly Emerging Diseases in a World Out of Balance* (New York: Penguin, 1994).

p. 138 "resistance is almost certain . . .": Ibid., p. 48, citing the United Nations report *Malaria Eradication: International Development Advisory Board, April 13, 1956.*

p. 139 Lice found on American and Malaysian heads: R. J. Pollack et al., "Differential Permethrin Susceptibility of Head Lice Sampled in the United States and Borneo," *Archives of Pediatrics and Adolescent Medicine* 153 (September 15, 1999): 969–73.

p. 140 Pesticides and agriculture are discussed in D. Pimentel and H. Lehman, eds., *The Pesticide Question: Environment, Economics and Ethics* (New York: Chapman and Hall, 1993).

p. 140 A good example of genetic evolution of mosquito resistance is in N. Pasteur and M. Raymond, "Insecticide Resistance Genes in Mosquitoes: Their Mutation, Migration, and Selection in Field Populations," *Journal of Heredity* 87 (1996): 444–49.

p. 145 Bt references include B. Tabashnik et al., "One Gene in Diamondback Moth Confers Resistance to Four *Bacillus thuringiensis* Toxins," *Proceedings of the National Academy of Science (USA)* 94:1640–44; and John A. McKenzie, *Ecological and Evolutionary Aspects of Insecticide Resistance* (Austin: R. G. Landes/Academic Press, 1996).

p. 149 On Bt resistance, see B. E. Tabashnik, "Evolution of Resistance to *Bacillus thuringiensis*," *Annual Review of Entomology* 39 (1994): 47–49.

p. 150 1.8 million acres: J. Kaiser, "Pests Overwhelm *Bt* Cotton Crop," *Science* 273 (1996): 423.

p. 151 Prior frequency of Bt resistance was measured by Y.-B. Liu et al., "Field-Evolved Resistance to *Bacillus thuringiensis* Toxin CryIC in Diamondback Moth (Lepidoptera: Plutellidae)," *Journal of Economic Entomology* 89 (1996): 798–804.

p. 152 Evolution in the face of multiple toxins was reported by G. P. Georghiou and M. C. Wirth, "Influence of Exposure to Single Versus Multiple Toxins of *Bacillus thuringiensis* subsp. *israelensis* on Development of Resistance in the Mosquito *Culex quinquefasciatus* (Diptera: Culicidae)," *Applied and Environmental Microbiology* 63 (1997): 1095–101.

p. 156 John Losey and colleagues described Bt effects on monarch butterflies in J. E. Losey, L. S. Raynor and M. E. Carter, "Transgenic Pollen Harms Monarch Larvae," *Nature* 399 (1999): 214. A more well controlled experiment on swallowtails recently appeared in C. L. Wraight et al., "Absence of Toxicity of Bacillus thuringiensis Pollen to Black Swallowtails Under Field Conditions," *Proceedings of the National Academy of Sciences (USA)* 97 (2000): 7700–7703.

p. 158 The refuge strategy was tested in Y.-B. Liu and B. E. Tabashnik, "Experimental Evidence That Refuges Delay Insect Adaptation to

Bacillus thuringiensis," Proceedings of the Royal Society of London, Series B, *Biological Science* 264 (1997): 605–10; and J. Mallet and P. Porter, "Preventing Insect Adaptation to Insect-Resistant Crops: Are Seed Mixtures or Refugia the Best Strategy?" Proceedings of the Royal Society of London, Series B, *Biological Science* 250 (1992): 165–69.

Chapter 7: Biotechnology and the Chemical Plow

p. 162 The authors had discovered: H. C. Steinruken and N. Amrhein, "The Herbicide Glyphosate Is a Potent Inhibitor of 5-Enopyruvyl-Shikimic Acid-3-Phosphate Synthase," *Biochemistry and Biophysical Research Communications* 94 (1980): 1207–12.

p. 164 Table 7.1: P. Westra, "Species Diversity and Geographic Range of Herbicide Resistance Weeds in the Western Region of North America," *Proceedings of the Western Society of Weed Science* 46 (1993): 121–22.

p. 165 For a description of one resistant weed, see "The Pigweeds: They're Prolific and Prone to Resistance," at the website of Crop-net.com, which is based in Eugene, Oregon. The URL is http://www.crop-net.com/archive/mwft196t.htm.

p. 166 A comprehensive compilation of facts about weed resistance is I. M. Heap, "The Occurrence of Herbicide-Resistant Weeds Worldwide," *Pesticide Science* 51 (November 1997): 235–43; or the article can be read on-line at http://www.weedscience.com/paper/resist97.htm.

p. 166 Kansas State Board of Education: In the summer of 2000, primary elections removed two key critics of science teaching from the school board ballot. This may herald a shift in Kansas policies.

p. 170 John Doebley's work on maize includes J. Doebley et al., "Genetic and Morphological Analysis of a Maize Teosinte F2 Population—Implications for the Origin of Maize," *Proceeding of the National Academy of Science (USA)* 87 (1990): 9888–92; and J. Doebley, "Molecular Evidence and the Evolution of Maize," *Economic Botany* 44 (1990): 6–27.

p. 171 To find out more about glyphosate toxicity and the field escape of resistant plants, see "Glyphosate Fact Sheet," Greenpeace Media Center, April 1997.

p. 172 Engineered mustard plants resistant to glyphosate: H. J. Klee, Y. M. Muskopf, and C. S. Gasser, "Cloning of an *Arabidopsis thaliana* Gene Encoding 5-enolpyruvylshikimate-3-phosphate Synthase: Sequence Analysis and Manipulation to Obtain Glyphosate-Tolerant Plants," *Molecular and General Genetics* 210 (1987): 437–42.

p. 174 Concerns about escape of genes from crops into weeds have been discussed regarding many agricultural systems. For wheat, see R. S. Zemetra, J. Hansen, and C. Mallory-Smith, "Potential for Gene Transfer Between Wheat (*Triticum aestivum*) and Jointed Goatgrass (*Aegilops cylindrica*)," *Weed Science* 46 (1998): 313–17. For canola, see M. Reiger, C. Preston, and S. Powles, "Risks of Gene Flow from Transgenic Herbicide-Resistant Canola (*Brassica napus*) to Weedy Relatives in Southern Australian Cropping Systems," *Australian Journal of Agricultural Research* 50 (1999): 115–28. For corn, see A. Abbo and B. Rubin, "Transgenic Crops: A Cautionary Tale," *Science* 287 (2000): 1927–28, and citations therein.

p. 175 See website "Jointed Goatgrass: A Threat to U.S. Wheat Production," on-line (last update March 11, 2000) at http://www.ianr.unl.edu/jgg/index.htm.

Chapter 8: Evolution All at Sea

A good general source is J. Burger et al., eds, *Protecting the Commons: A Framework for Resource Management in the Americas* (Washington, D.C.: Shearwater Books, 2000).

p. 186 The life of cod and cod fisheries is engagingly reported by M. Kurlansky, *Cod: A Biography of the Fish That Changed the World* (New York: Penguin, 1998).

p. 187 Selection size in salmon is chronicled in W. E. Ricker, "Changes in the Average Size and Average Age of Pacific Salmon," *Canadian Journal of Fisheries and Aquatic Science* 38 (1981): 1636–56. Figure 8.1 is redrawn from an illustration in ibid.

p. 188 Changes in lake whitefish populations were given an evolutionary explanation by P. Handford, G. Bell, and T. Reimchen, "A Gillnet Fishery Considered as an Experiment in Artificial Selection," *Journal of Fisheries Research Board of Canada* 34 (1977): 954–61.

p. 193 The two strategies had about equal chance: M. Gross, "Salmon Breeding Behavior and Life History Evolution in Changing Environments," *Ecology* 72 (1991): 1180–86.

p. 195 For sex change in shrimp, see E. L. Charnov, "Sex Reversal in *Pandalus borealis*: Effect of a Shrimp Fishery?" *Marine Biology Letters* 2 (1981): 53–57.

p. 197 The classic evolutionary view of change in snail shape is in R. H. Seeley, "Intense Natural Selection Caused a Rapid Morphological Change in a Living Marine Snail," *Proceedings of the National Academy of Science (USA)* 83 (1986): 6897–901. But a different view has been proposed by G. C. Trussel, "Phenotypic Plasticity in an Intertidal Snail: The Role of a Common Crab Predator," *Evolution* 50 (1996): 448–54; and G. C. Trussel and L. D. Smith, "Induced Defenses in Response to an Invading Crab Predator: An Explanation of Historical and Geographic Phenotypic Change," *Proceedings of the National Academy of Science (USA)* 97 (2000): 2123–27.

p. 199 Warner's work on sex change in general, and blue-headed wrasses in particular, is legendary. For a recent treatment, see Warner et al., "Sexual Conflict: Males with Highest Mating Success Convey the Lowest Fertilization Benefits to Females," Proceedings of the Royal Society of London, Series B, *Biological Sciences* 262 (1995): 135–39. See also D. R. Robertson, "Social Control of Sex-Reversal in a Coral Reef Fish," *Science* 177 (1972): 1007–9.

p. 201 For some of the best shark stories, see W. D. Westervelt, *Myths and Legends of Hawaii* (Honolulu: Mutual Publishing, 1989).

p. 202 Lou Herman describes the idea that humpback whales are recent additions to Hawaii in L. M. Herman, "Humpback Whales in Hawaiian Waters: A Study in Historical Ecology," *Pacific Science* 33 (1980): 1–15.

Chapter 9: Are Humans Still Evolving?

Overall sources about human evolution are P. Erlich, *Human Natures* (New York: Island Press, 2000), and C. Wills, *Children of Prometheus: The Accelerating Pace of Human Evolution* (Reading, Mass.: Perseus Books, 1998).

p. 210 The relationship between milk diets, livestock husbandry, and lactose tolerance is described by C. Holden and R. Mace, "Phylogenetic Analysis of the Evolution of Lactose Digestion in Adults," *Human Biology* 69:605–28.

p. 211 "Selection pressures have actually shifted . . .": C. Wills, *Children of Prometheus: The Accelerating Pace of Human Evolution* (Reading, Mass.: Perseus Books, 1998), p. 25.

p. 212 Figures on infant mortality and health spending in Africa are from L. Garrett, *The Coming Plague: Newly Emerging Diseases in a World Out of Balance* (New York: Penguin, 1994), pp. 205–6.

p. 212 Health expenditures are available from the U.S. Health Care Financing Administration (hcfa.gov/stats) and *Bulletin of the World Health Organization* 78 (2000): 1–152.

p. 212 "The best remedy . . .": Quoted in F. Braudel, *Civilization and Capitalism: Fifteenth to Eighteenth Century*, vol. 1, *The Structures of Everyday Life: The Limits of the Possible*, trans. S. Reynolds (Berkeley, Calif.: University of California Press, 1992), p. 81.

p. 215 For cystic fibrosis, a description of its symptoms, and genetics, see S. Gabriel et al., "Cystic Fibrosis Heterozygote Resistance to Cholera Toxin in the Cystic Fibrosis Mouse Model," *Science* 266 (1994): 107–9; and N. Morral et al., "The Origin of the Major Cystic-Fibrosis Mutation (ΔF508) in European Populations," *Nature Genetics* 7 (1994): 169–75.

p. 220 "Until only a few hundred years ago . . .": P. M. Quinton, "What Is Good About Cystic Fibrosis?" *Current Biology* 4 (1994): 742–43.

p. 221 For AIDS resistance and genetics, see S. J. O'Brien, "AIDS: A Role for Host Genes," *Hospital Practice*, July 15, 1998, pp. 53–79; S. J. O'Brien and M. Dean, "In Search of AIDS-Resistance Genes," *Scientific American* 277 (1997): 44–51; and J. C. Stephens et al., "Dating the Origin of the CCR5-Δ32 AIDS-Resistance Allele by the Coalescence of Haplotypes," *American Journal of Human Genetics* 62 (1988): 1507–15.

p. 229 Wealth statistics are from E. Kapstein, *Sharing the Wealth: Workers and the World Economy* (New York: Norton, 1999); and P. Krusell and A. A. Smith, "Income and Wealth Heterogeneity in the Macroeconomy," *Journal of Political Economy* 106 (1998): 867–96.

Chapter 10: The Ecology and Evolution of Aloha

There are many books on human culture. See especially S. Pinker, *How the Mind Works* (New York: Norton, 1997), J. Diamond, *The Third Chimpanzee* (New York: HarperCollins, 1992), and P. Ehrlich, *Human Natures* (New York: Island Press, 2000).

p. 234 "Nothing makes noble persons . . .": F. Braudel, *Civilization and Capitalism: Fifteenth to Eighteenth Century*, vol. 1, *The Structures of Everyday Life: The Limits of the Possible*, trans. S. Reynolds. (Berkeley, Calif.: University of California Press, 1992), p. 324.

p. 235 "incredibly slow phase . . .": R. Lewin, *Human Evolution: An Illustrated Introduction*, 4th ed. (Malden, Mass.: Blackwell, 1999), pp. 197–98.

p. 236 For Thomas Huxley, see Adrian Desmond, *Huxley: From Devil's Disciple to Evolution's High Priest* (Reading, Mass.: Addison-Wesley, 1997).

p. 239 Finches of the Galápagos Islands: B. R. Grant and P. R. Grant, "Cultural Inheritance of Song and Its Role in the Evolution of Darwin's Finches," *Evolution* 50 (1996): 2471–87.

p. 243 For a recent popularization of memes, see S. Blackmore, *The Meme Machine* (New York: Oxford University Press, 2000).

p. 244 "The common designation of 'evolution' . . .": S. J. Gould, *Full House: The Spread of Excellence from Plato to Darwin* (New York: Three Rivers Press, 1997), p. 219.

p. 244 "I think you'll agree . . .": S. Pinker, *How the Mind Works* (New York: Norton, 1997), p. 209.

p. 245 "each mind has a limited capacity for memes": D. Dennett, *Darwin's Dangerous Idea* (New York: Simon & Schuster, 1995), p. 349.

p. 247 The history and solution of Fermat's theorem is detailed in A. D. Aczel, *Fermat's Last Theorem: Unlocking the Secret of an Ancient Mathematical Problem* (New York: Doubleday, 1997).

Index